A Century of Materials Science and Engineering at Stanford

From Steels to Semiconductors to Nano- and Bio-Materials

William D. Nix
Lee Otterson Professor of Engineering, Emeritus

Stanford Historical Society
and
The Department of Materials Science and Engineering
Stanford University

2019

Stanford | 1919–2019
Materials Science & Engineering

This project, honoring the 100th anniversary of the MSE Department, was made possible through funding from the Materials Science & Engineering Department, Stanford University

Written by Professor William D. Nix
Edited by Roxanne L. Nilan and Jean M. Deken
Format production by Roxanne Nilan and Ralph Simpson

Stanford Historical Society
P.O. Box 20028
Stanford, California 94309
(650) 725-3332
historicalsociety.stanford.edu

ISBN 978-0-9847958-4-0

COVER PHOTO: *The Mining and Metallurgy Building, longtime former home of Materials Science & Engineering, is seen here sometime after 1949. That year, the courtyard (the low, light-shaded area between the arms of the U) was covered to create more lab space. (Stanford University Archives)*

ENDNOTES: *Parenthetical word and date at end of a direct quote signals an endnote (pp. 235-36).*

PHOTO CREDIT: *Follows each caption in parentheses.*

Table of Contents

Preface i

Introduction 1

Part I ~ Beginnings

Chapter 1 ~ Early Teaching of Materials (Metallurgy) at Stanford 3

Chapter 2 ~ Theodore J. Hoover and the Founding of the Department of Mining and Metallurgy 9

Part II ~ From Metallurgy to Materials Science

Chapter 3 ~ Welton J. Crook: Steel Metallurgist, Par Excellence 19

Chapter 4 ~ O. Cutler Shepard: From Assayer of Gold and Silver To Head of the Materials Science Department 25

Chapter 5 ~ R.A. Huggins and the Creation of Materials Science at Stanford 35

Part III ~ Building a National Reputation

Chapter 6 ~ Nucleation of a Nationally Prominent Department 45

Chapter 7 ~ Appointments of Future Department Chairs 57

Chapter 8 ~ Appointments in Metallurgy and Failure Analysis 71

Chapter 9 ~ Completing the Faculty Expansion 81

Chapter 10 ~ Visiting Professors and Lecturers and the Intimate Nature of the Department of the 1960s 97

Chapter 11 ~ Professional Society Connections 105

Chapter 12 ~ Women in Materials Science 109

Chapter 13 ~ Student-Faculty Sports Activities 115

Chapter 14 ~ International Students 123

Chapter 15 ~ Broadening the Faculty in the 1970s 129

Chapter 16 ~ Professors (Research) 137

Chapter 17 ~ Joint Faculty Appointments 145

Chapter 18 ~ Consulting Professors: Enriching the Teaching and Research Programs in the Department 157

Chapter 19 ~ Faculty Appointments of the 1980s 163

Chapter 20 ~ Department Leadership and Troubles for an Aging Department 175

Part IV ~ The Modern Department

Chapter 21 ~ Rebuilding for the 21st Century ~ The Beginning 181

Chapter 22 ~ Rebuilding for the 21st Century ~ The Explosion 191

Chapter 23 ~ Appointments for Energy and the Environment 205

Chapter 24 ~ Appointments in Biomaterials 211

Chapter 25 ~ Appointments in Materials Physics 219

Part V ~ Closing Observations 227

List of Acronyms and Abbreviations 231

Endnotes 235

Bibliography 237

Name Index 240

Preface

The origins of the Department of Materials Science and Engineering at Stanford trace back to early coursework in mining, but ultimately to the founding of the Mining and Metallurgy Department in 1919. Thus, in 2019, the department celebrates the 100th anniversary of its founding. What better time to look back and assess the progress that has been made in the study of materials at Stanford since the early part of the 20th century.

This documentary history has only the barest of overarching themes guiding this story. One is an account of the various and changing types of materials that have drawn the attention of the department's faculty and students over the past century, as reflected in the title of this work. Interest in different kinds of materials at different times has been in response to the ever-changing needs of industry and society. Another theme involves the inevitable triumphs and tribulations, opportunities and challenges, that this small department has experienced in a highly competetive and ambitious university environment. The main objective, however, is simply to record and document events and people who have made up the history of the department.

Divided into five parts, it follows a chronological order. Part I, "Beginnings," describes the pre-history of the department, that is, the people and events in metallurgy at Stanford before the department's establishment, and the circumstances behind its founding in 1919. Part II, "From Metallurgy to Materials Science," follows three key figures whose careers illuminate the evolution of the Mining and Metallurgy Department of the 1920s into a Materials Science Department in 1960. Part III, "Building a National Reputation," describes the rapid growth and maturation of the department that took place from the 1960s to the 1990s. And finally, Part IV, "The Modern Department," describes the rebirth of the department and its explosive growth in the early part of the 21st century. Finally, "Closing Observations" suggest some of the sweeping changes that have, and in one case have not, occurred in the study of materials at Stanford over the past century.

Having been in the department for 60 years, I have known many of the people who have defined its history, including some who started their careers here in the early part of the 20th century. Much of what I have written is based on my knowledge and memory of what has transpired here, as well as on various university documents and publications I have consulted to verify places, people, and dates and bring context to the story. Two informal histories, written by faculty colleagues, have been especially helpful. In 1968, Professor Welton J. Crook presented "A Partial History of Metallurgy at Stanford University" on the occasion of Professor O. Cutler Shepard's retirement. It was an invaluable, if highly personal, history of mining and metallurgy at Stanford from the 1890s to about 1940 by a man who had seen much of that early history. Crook brings color to the story, with an emphasis on student life, teaching and research, and on many of the faculty who served as his colleagues. Professor Shepard's "Historical Sketch of the Department of Metallurgy and Materials Science at Stanford," was printed by the department in 1992 to celebrate his 90th birthday. Shepard was in a good position to review the evolution of the department from the 1920s to the 1960s, from metallurgy to materials science, having served on the faculty for more than 40 years.

I have kept essentially all the minutes of the MSE faculty since 1957 in my personal files. I have also collected many of key documents pertaining to the history of the department for many decades, including those mentioned above. It is my intention to donate these documents to the Stanford University Archives so that they may be available to other interested persons. Also available in the University Archives are the extensive records of the School of Engineering, as well as the papers of Stanford's presidents and provosts.

I am grateful to university departments and libraries for granting me access to many useful archival and photographic collections, especially the Stanford University Archives and the Stanford News Service. In addition to many historical Stanford publications available there, our student newspaper, the *Stanford Daily*, now searchable online, was especially helpful in finding information about early faculty, students, and alumni. Also available (and searchable) online are the following invaluable digitized university publications: the *Annual Register* (1891-1947) (today's Courses and Degrees), the *Annual Report of the President* (1903-1948), faculty *Memorial Resolutions of the Academic Council*, and the Stanford Alumni Association's monthly magazine (which has had many names).

In putting this retrospective together, I interviewed several faculty members who have retired, and I asked all current faculty members to provide information to me. From these, I created biographical sketches about them and their careers at Stanford. Although I edited each of the sketches submitted, I hope individual personalities still show through.

Many faculty members and students have generously shared stories, photographs, and documents that have been invaluable in compiling this story. They include: Ralph Parkman, William Tiller, John Shyne, Alan Ardell, David Barnett, Robert Feigelson, Alan Miller, Arthur Bienenstock, Robert Huggins, Paul Flinn, Darcy Hughes, Alice Fischer-Colbrie, Clayton Bates, Robert Sinclair, Robert L. White, Theodore Geballe, Friedrich Prinz, Andrew Spakowitz, Jeffrey Wadsworth, Michael Kelly, Seung-Min Han, W. Robert Johnson, Toshi Oyama, Michael Mills, John Bravman, Bruce Clemens, Shan Wang, Reinhold Dauskardt, Paul McIntyre, Michael McGehee, Mark Brongersma, Nicholas Melosh, Yi Cui, Alberto Salleo, Sarah Heilshorn, Aaron Lindenberg, Jennifer Dionne, Evan Reed, William Cheuh, Eric Appel and Guosong Hong.

In addition, I appreciate the suggestions of several people who have read parts of this manuscript: Jean A. Nix, David M. Barnett, Ralph Parkman, and Bruce Clemens.

Finally, I gratefully acknowledge the financial support for this publication by the Department of Materials Science and Engineering, and the enthusiastic support and encouragement of MSE chair Alberto Salleo and former chair Paul C. McIntyre, and the MSE staff. I also wish to acknowledge the Stanford Historical Society's Publications Committee, and especially Dr. Roxanne ("Rocky") Nilan for guiding me in this work and for bringing it to a successful conclusion. Rocky brought her amazing knowledge and passion about Stanford history, as well as her special skills as a historian, editor, and former Stanford University Archivist, to greatly improve this work. Her contributions are much appreciated. I also wish to thank Jean Deken, Ralph Simpson and Karen Bartholomew for their expert assistance in helping to create this polished document.

William D. Nix,
Stanford, California, 2019

Left: Cutler Shepard instructs a student in optical microscopy, 1950s. (Stanford University Archives)

Below: The latest optical microscope, 1957, for metallurgical engineering. (Stanford University Archives)

Above: Electron microprobe lab, Center for Materials Research, 1980s. (Stanford News Service)

Right: Electron microprobe, 1980. (Stanford News Service)

Materials Science lab (above) and Materials Science machine shop (below) in the Peterson Lab Building in the 1970s. (Stanford News Service)

Introduction

The study of materials at Stanford started in its mining curriculum of the 1890s, which was, of course, focused on the extraction and processing of metals from ores. As that field developed, more attention was placed on the thermal and mechanical processing and properties of metals. This, in turn, led to the development of techniques for studying the internal structure of materials and the relation of microstructure to properties. Through this evolution, the field of materials science and engineering was born.

Initially drawn out of the Geology and Mining Department in 1919, the new Department of Mining and Metallurgy, because its origins were in mining, focused mostly on metals for nearly half a century. Many other departments of materials science and engineering, particularly those in the American West—a resource-rich geography demanding well-trained engineers and managers—had a similar history and connection to mining. The Department of Materials Science and Engineering at U.C. Berkeley, for example, grew out of a College of Mines and emphasized metals for the first half of the 20th century. Indeed, that department still is partly housed in the Hearst Memorial Mining Building on the Berkeley Campus, a building dedicated to the memory of the successful mining mogul George Hearst. The Materials Department at the University of Washington also originated in a School of Mining Engineering, in this case influenced by the great Yukon gold discoveries, but that department, like many others, eventually adopted the broader mission of materials science and engineering. Other schools, such as the Colorado School of Mines, retained their long-standing association with mining in their names, even though they, too, adopted a broad vision of materials in their materials department. The trajectory of the Colorado School of Mines must have been what Theodore J. Hoover had in mind when, in accepting the position as executive head of the Department of Mining and Metallurgy, he wrote: "Stanford University should have the greatest mining school in the world." Other schools, like the South Dakota School of Mines, altered their names slightly (South Dakota School of Mines and Technology) while still others, like the Missouri School of Mines, made more major changes in their names (Missouri University of Science and Technology) reflecting changes in focus. By comparison to these western schools, the path of the Materials Department at the University of Michigan is seemingly unusual, as it grew out of chemistry and chemical engineering, and maintained this focus in the early part of the 20th century. Interestingly, our MSE Department has a prehistory in courses taught by Stanford chemists.

Much of the department's attention through the 1920s and 1930s was on steels and their compositions, microstructures, and properties. But by the 1940s and 1950s, with connections to the mining industry having receded into the distant past, students and faculty in the department were addressing the much wider materials issues associated with technologies of the day, such as nuclear power and aerospace applications. By then the study of metals such as aluminum, magnesium and titanium alloys had been brought under the emerging materials science umbrella. Major changes were to occur in the 1960s and 1970s, with the development of microelectronics, as attention was directed to silicon and other semiconductors, which are central to microelectronic technology. Work on electro-chemical materials that would have applications for batteries and fuel cells was also initiated. By the beginning of the 21st century, with work continuing on materials for microelectronic and electro-chemical device applications, attention was drawn to the role that materials science plays in the broad fields of nano-sciences and biosciences. Thus, the underlying theme throughout this department history will be a focus on those materials that enable the technologies of the day, now including the environment, biology, and medicine.

William F. Durand Building, current home of the Materials Science & Engineering Department. (MSE Dept.)

Part I ~ Beginnings

Chapter 1

Early Teaching of Materials (Metallurgy) at Stanford

As we celebrate the centennial of the Department of Materials Science and Engineering, it is worth surveying the department's prehistory, from a twinkle in the eye of an influential geology professor in 1891 to the university's first metallurgists and its prototype course in materials science.

Courses about metallurgy and assaying were listed (though not taught) in the first university bulletin of newly opened Leland Stanford Junior University in 1891. Over the next ten years, Stanford would rethink its approach to mining and metallurgy with a handful of courses taught within the departments of Geology, Chemistry, Electrical Engineering, and Civil Engineering. Its first metallurgy course, taught by Stanford's first "professor of mining and metallurgy," would appear in 1899, but it would not be until 1903 that the first true course in the study of materials appeared, taught by chemist Dorsey Lyon in the newly expanded Geology and Mining Department.

Prehistory in Geology and Chemistry

At the University's opening, in 1891, the Geology Department had been among its 16 original departments and was one of its most influential. John Casper Branner, its executive head, had been the first professor hired by President David Starr Jordan. Branner, a close friend of Jordan and one of his most trusted advisors, quickly became a powerful voice in campus affairs. (He would serve as Jordan's vice-president, and still later, as Stanford's second president.) As state geologist of Arkansas, Branner had become expert in detailed, well-written geological surveys—a skill he would hone in Brazil as well as in the American West—and he brought to Stanford a firm belief in the practicality of academic research and the need to supply the mining industry with both accurate information and well-informed, scientifically-trained mine managers. Among his first students was Herbert Hoover, of Stanford's pioneer class of 1895, who would become an internationally successful mining manager and investor by the time he was 40. Hoover, and his brother, Theodore, another Branner student, exemplified the kind of successful mining expert Stanford expected to produce.

As Jordan and other academic advisors to university founders Leland and Jane Stanford created a tentative curriculum for the new university, 16 departments took shape to welcome a full round of undergraduate and graduate students in the fall of 1891. Eight additional departments were rather speculatively listed in the course bulletin, among them a tentative "Mining and Metallurgy Department," with five courses (to be taught no earlier than 1893) with a daunting list of course prerequisites for the five brave students who would list this as their major that year. (A year later, the "Metallurgy" disappeared from the name.)

No faculty was listed for the tentative department, and the one course to get underway, in 1894, "Assaying," would be taught by chemistry professor Lionel Remond Lenox. By this time, the university had given up on establishing a separate department, wrapping the mining courses into the Geology Department. In 1897, that department was renamed Geology and Mining.

Lenox was one of those young pioneering professors who, at the breakout point of their career, Jordan could convince to take a chance on Stanford. The 27-year old had worked in industry (two iron manufacturing companies) and with the U.S. Navy after receiving his Columbia Ph.D. in chemistry (1888). He arrived at Stanford in 1897 as an associate professor of analytical chemistry (he was soon promoted to professor). He became an expert on qualitative analysis of iron and steel in a department focused on industrial chemistry. Lenox continued to teach assaying for mining and metallurgy students well into the 1920s.

John F. Newsom (Stanford University Archives)

John F. Newsom ~ First Professor of Mining and Metallurgy

With Branner's strong connections with the U.S. Geological Survey and with Western mining companies, the Geology and Mining Department's coursework remained focused on finding, identifying, and extracting mineral resources. In 1898, its two professors (Branner, geology, and J. P. Smith, paleontology) welcomed John Flesher Newsom as Stanford's first "professor of mining and metallurgy." Newsom, an Indiana native, had studied with Branner at both Indiana University (A.B., 1891) and Stanford (A.M., 1893), meanwhile serving as Branner's assistant on the Arkansas State Geological Survey. (He named his son John Branner Newsom.) Newsom became a full professor on receiving his Stanford Ph.D. in 1901.

Newsom taught the Department's first metallurgy class, with "special reference to the metallurgy of copper, lead, silver and gold," but he soon moved on to focus on economic geology, field geology, and mining operations after Dorsey Lyon arrived at Stanford in 1903. In 1918, Newsom left Stanford (he remained a Palo Alto resident until 1926) to work for the Yukon Gold Company of New York in Kuala Lumpur. He died in 1928, aged 59, in the midst of a successful career as an international mining consultant.

Dorsey A. Lyon ~ The First Course in Materials at Stanford

It can be argued that the very first course in materials science and engineering at Stanford was given in the academic year 1903-04 by Dorsey Alfred Lyon, an instructor in the department of Geology and Mining. In that year, and for a few years thereafter, Lyon offered a course entitled, "Metallography and Physics of the Metals." In his course description, Lyon included the following prophetic lines: "…work will include a study of the microstructure of the industrial metals and alloys, and of the influence of chemical composition, thermal and mechanical treatment, upon their structure. Also, the relation of the structure of metals and alloys to their physical properties."

Dorsey A. Lyon, 1898. (Stanford University Archives)

One could hardly have created a better working definition of the subject of materials science for most of the 20th century.

Dorsey Lyon had graduated from Stanford with an A.B. degree in chemistry in 1898. (The A.B. stood for "Artium Baccalaureus"; it was later called the Bachelor of Arts (B.A.) degree.) That year, the 27-year old Lyon joined the University of Washington's fledgling School of Mining Engineering as Seattle's mining industry struggled to keep up with new discoveries in the Yukon and Alaska gold fields. In three years, the versatile Lyon became assistant professor of geology and physical geography, an assistant professor of mining engineering, and an instructor in chemistry. By 1899, he had developed courses in mining and metallurgy, including mineral formation, mining surveys, mine development, mining machinery, assaying, ore testing and smelting. In 1901, Lyon, now a full professor, was offered the post of dean of Washington's new School of Mines, but he left Washington for Harvard before the appointment became official. It has been suggested that the university's contradicting efforts to create both a school of mines, or mining engineering, and a college of engineering with a mining engineering department, encouraged him to move on.

While doing post-graduate work in metallurgy at Harvard, Lyon studied with Professor Albert Sauveur, one of the founders of the modern science of metallurgy. Sauveur had developed the first university laboratory of metallurgy. It is almost certain that the course Lyon offered at Stanford was patterned after Sauveur's courses. (The title of Sauveur's seminal 1912 textbook, *The Metallography and the Heat Treatment of Iron and Steel*, includes some of the terms used in Lyon's course description.) Lyon, like Sauveur, was also influenced by the writings of Columbia University's Henry Marion Howe. Lyon mentioned in his course description that he would be following Howe's famous textbook, *The Metallurgy of Steel* (1891). As an interesting aside, Howe, later known as the dean of American metallurgists, was the only son of prominent social activist Julia Ward Howe, author of "The Battle Hymn of the

Republic," and Samuel Gridley Howe, a prominent physician, abolitionist, and advocate of education for the blind.

After receiving his M.S. degree from Harvard in 1902, Lyon accepted a position to assist in the construction of a plant in Utah for the U.S. Mining and Smelting Co., which was then developing mines throughout the Western United States and Mexico. Just a year later, he returned to Stanford on a temporary basis to fill a vacancy on the faculty as an instructor. His stay proved longer than he had intended, lasting from 1903 to 1907. Lyon left to develop electro-furnace smelting of iron ore in Shasta County. After a year back at Harvard, he joined the U.S. Bureau of Mines. In 1917, he was appointed supervisor of the Bureau's 13 mining experimental stations across the country, and two years later was made its chief metallurgist. He retired to Palo Alto in the late 1930s and died there in 1945.

The course Lyon created, "Metallography and Physics of the Metals," disappeared from the curriculum after 1907; a course with that description did not appear again until the late 1940s.

Galen H. Clevenger, 1920. (Engineering and Mining Journal, 1920)

Galen H. Clevenger ~ Bringing Sponsored Research to Teaching

Following Lyon's departure from Stanford, courses on physical metallurgy were taught by Galen Howell Clevenger, who had come to Stanford in 1905 after receiving degrees from South Dakota School of Mines (1901) and Columbia University (1903). Like Lyon, he was considered not only a very able metallurgist and, at least initially, an enthusiastic teacher. A difficult job market in the mining industry during the first decade of the new century took a toll on the number of students drawn to Stanford's department, but metallurgy began to revive and even flourish during the following decade. Mining and metallurgy, closely entwined, had become a distinct division of the Geology and Mining Department by 1910. Clevenger and Luther Bahney (who came with copper and lead mining experience) covered metallurgy for the growing department, with the assistance of a young Australian student, Welton Crook.

Clevenger, a gold and silver cyanidation expert, became a very prominent metallurgist and was perhaps the first faculty member at Stanford to have a sponsored research project in materials. His research project, sponsored by the U.S. Smelting, Refining and Mining Company, involved an effort to discover the causes of failure of a large zinc plant. He discovered that the presence of cobalt in the ore led to hydrogen being created in the smelting process rather than zinc, and he devised a method for removing the troublesome cobalt from the ore before the extraction of zinc. This work by Clevenger also foretold the contributions that Stanford faculty would make to industry through their research.

During his last years at Stanford, Clevenger took the title "research professor," and spent much of his time on his industry consulting and research, often drawing in his Stanford students. In 1918, Clevenger was among the many Stanford faculty to take up a government post during the First World War, serving as a metallurgist in charge of precious metals investigations for the U.S. Bureau of Mines. He subsequently left the faculty to become head of research for the U.S. Smelting, Refining and Mining Company, and later went on to become president of that company. He died in 1953.

Waldemar F. Dietrich

Waldemar Fenn Dietrich taught process metallurgy, ceramics, and mining at Stanford from 1916 to 1930, and played a role in the founding of the Mining and Metallurgy Department.

Dietrich was born in Portland, Oregon, in 1892, and attended Stanford, receiving his A.B. degree in geology in 1913 and his Engineer's degree in 1914. He worked briefly in Arizona before starting as a lecturer in the Mining and Metallurgy division in the Geology and Mining Department in 1916. He was quickly promoted to assistant professor in 1917 and to associate professor a year later, at the age of 26, making up for Clevenger's gradual disappearance from the teaching ranks. Dietrich continued to teach courses in process metallurgy and mining until he resigned in 1930. He was appointed professor in his last year on the faculty.

In 1928, Dietrich wrote a major report entitled "The Clay Resources and the Ceramic Industry of California," published by the State Division of Mines and Mining of the California Department of Natural Resources. His close association with the Division of Mines and Mining might have been the reason for his resignation from the Stanford faculty in 1930

Waldemar F. Dietrich, 1956. (Stanford News Service)

and his move to Sacramento. He was listed as a Teacher of Mining and Metallurgy at Sacramento Junior College (the future Sacramento State University) in 1940, when he co-authored the textbook *Fire Assaying* with O. Cutler Shepard, who had replaced him at Stanford.

During the 1950s, he served as chief of the Ceramic and Fertilizer Division of the U.S. Bureau of Mines. He died in Tuscon, Arizona, in 1980.

Electric furnace built by Galen Clevenger and Dorsey Lyon in 1908.
(Stanford University Archives)

Chapter 2

Theodore J. Hoover and the Founding of the Department of Mining and Metallurgy

In the years following World War I, the growing strength of mining and metallurgy in the Geology and Mining Department finally led to the creation of a department that had been envisioned at the university's opening, nearly 30 years earlier. The Department of Mining and Metallurgy, created in 1919, can be regarded as the founding department that ultimately led to today's Department of Materials Science and Engineering.

It is highly likely that a split department was not what the Geology and Mining Department's executive head, Bailey Willis, had in mind when he consulted the university's new president, Ray Lyman Wilbur, about expanding mining, metallurgy, and petroleum engineering within his present department. By the end of American involvement in the European war, his department had lost three of its four mining and metallurgy faculty, with only Waldemar Dietrich still on board. It clearly was time to rebuild.

Wilbur turned to Theodore J. Hoover, a successful mining manager, consultant, and Stanford alumnus, for advice. The result was a new department, with Theodore Hoover as its head, dramatic curriculum changes, expansion of the Engineer's degree, and, ultimately, in 1925, a new School of Engineering.

Theodore J. Hoover ~ Mining Engineer and First Department Head

In recommending the creation of a department of mining and metallurgy, Theodore Hoover had very definite ideas on the purpose of engineering education and its relationship to industry. Recognizing that Stanford had limited resources, he believed that the university should not turn out yet more narrowly-trained men for specialized employment, but rather a smaller number of engineers of broader outlook who would take on industry leadership roles backed by broad training in science, economics, world affairs, history, and languages. Hoover favored a four-year unspecialized undergraduate program for the A.B. degree in engineering (pre-mining), followed by two years of graduate study leading to the Degree of Engineer.

While Hoover's stress on the Engineer's degree was strongly supported by his fellow engineering department heads, he also had less popular opinions about engineering students and faculty. In 1923, he complained to the president of the "superabundance of dirty cords, lack of collars, and general slouchiness" among engineering students that suggested a lack of professional pride. Hoover also favored drawing his faculty exclusively, when possible, from industry rather than academia, with or without graduate degree credentials. While he supported faculty research, he veered away from the more traditional science-based research (to aid resource development) of his former geology professors, preferring to encourage targeted research supported and prioritized by specific company funding as a means to bolster advances in the mining industry itself. *(Hoover, 1925)*

Hoover's strong opinions were drawn from his family background and personality as much as from his industry experience. Theodore "Tad" Hoover had followed his intense and ambitious younger brother Herbert to Palo Alto in 1895, where he worked while "Bert" (with a small loan from his big brother) was finishing his studies in geology and mining.

Theodore Hoover had been born in West Branch, Iowa, on January 28, 1871, the eldest of three children of Jesse Hoover, a blacksmith and dealer in agricultural equipment, and Huldah Minthorn Hoover, a teacher and Quaker elder. Orphaned as children, the siblings would often be separated as they were raised by various relations. Joining his younger brother in Oregon in 1887, Tad began studies at Friends Pacific Academy in Newberg (although he lived with relatives, he worked for his board). He left, however, when his stern relatives disapproved of his attending a dance. He then attended William Penn College (Friends) in Iowa, interspersed with work at a variety of jobs. By age 26, he had saved enough to return to college.

Enrolling at Stanford in 1897, Theodore Hoover also chose the Geology Department. With some financial support from his brother, who was already a successful mining manager, he graduated in 1901 at the age of 30. Perhaps as an older undergraduate, and recently married, he disapproved of the tradition of rough dress among engineering and geology students, even though this was considered symbolic of the university's democratic and anti-elitist founding ideal (an ideal the brothers otherwise strongly supported). While they were students, both he and young Herbert had leaned towards proper collar and tie.

Theodore J. Hoover (Stanford University Archives)

After a year as an assayer for the Keystone Consolidated Mining Company in California's Sierra Nevada foothills, Theodore became assistant manager for the Standard Consolidated Mine in Bodie, California, on the dry, forbidding plains east of the Sierras. He was promoted to manager of the operation just a year later. By 1906, he was a consulting engineer for the Cosmopolitan Proprietary Company Ltd, of London and Mexico, and a year later became general manager of Minerals Separation, Ltd., a London company with world-wide mining interests. Hoover was actively involved in development of the flotation concentration process, which made mining of low-grade ore financially viable, and he later authored one of the first books on the concentration of ores by flotation.

Following his younger brother's example, after four years with Minerals Separation Ltd., Theodore Hoover entered private practice as a consulting mining and metallurgical engineer with offices in London and San Francisco. His business prospered during this decade of aggressive mining investment, and he simultaneously held various positions as consulting engineer, managing director, director, and president of mining companies in Australia, Mexico, Burma, Finland, and Russia. But the overheated mining investment market, the economic ravages of world-wide war and post-war recession, and the Russian Revolution took a toll on the Hoover brothers' investments. During World War I, Theodore helped his brother's humanitarian work with the Commission for Belgium Relief and at the war's end, he returned to California, his San Francisco office, and his sea-side Rancho del Oso, near Pescadero (purchased in 1914), where he could enjoy his life-long love of the outdoors and interest in conservation.

He also returned to the nearby Stanford scene as something more than an interested alumnus. Theodore Hoover, now 47, easily fit into Wilbur's scheme to upgrade both faculty and curriculum, but he also brought special leverage. Brother Herbert Hoover, Stanford's most conspicuously successful alumnus, had been named to Stanford's Board of Trustees in 1912. Herbert had made an immediate impact, streamlining the board's organization and modernizing its financial procedures. Herbert's best friend was his classmate, the former Stanford dean of medicine, and recently appointed university president Ray Lyman Wilbur, '96. Since taking office in 1916, President Wilbur had brought new energy to Stanford's administration, updating its administrative and academic structure, recruiting new faculty, and upgrading research funding through his assiduous work with, among others, the Carnegie and Rockefeller foundations.

Already at work enhancing the medical, law, and education departments (each became a school), Wilbur had taken seriously Bailey Willis's suggestion about upgrading mining and metallurgy and pressed further. He asked Waldemar F. Dietrich, associate professor in the Geology and Mining Department, and the only mining and metallurgy faculty member left in the department in late 1918, to lay out the case for a stand-alone Mining and Metallurgy Department, altogether separate from the Department of Geology. Dietrich's January 4, 1919 letter in support of the idea was accompanied by a 30-page report of recommended curriculum reforms.

Given Wilbur's close relations with the Hoover brothers, it was natural for him to then consult Theodore Hoover, sending him Dietrich's recommendations. Hoover endorsed the separate department idea, feeling strongly that while mining and metallurgy had connections to geology, those subjects would be better served within their own structure. Thus, the Mining and Metallurgy Department was created, in 1919, with Theodore Hoover as its head. This was the department that eventually became today's Materials Science and Engineering Department.

A preview of Hoover's style as department head and, later, dean of engineering, can be seen in his three-page February 1919 letter to Wilbur accepting appointment as professor of mining and metallurgy and executive head of the department (at an annual salary of $7,200). In it, he denigrates Dietrich's curriculum recommendations as limited in ambition, and has even less praise for Stanford's competition. Columbia's School of Mines, was "wallowing in a fog of materialism and swamped with physical equipment upon which a pupil's time is wasted," he said. Another mining school he viewed as "decadent," and also noted that Britain's Royal School of Mines "will not shake off the incubus of the caste system in England

The Mining and Metallurgy Department and Labs, ca. 1935. Recently constructed Lagunita Court, opened in 1934, can be seen at top. (Stanford University Archives)

for generations." Hoover would rid the department of equipment and discourage manual training, while encouraging both faculty consulting for mining and petroleum interests and personal profit from patents resulting from on-campus research. *(Hoover, 1919)*

The Degree of Engineer

Creation of the Mining and Metallurgy Department in 1919 began a century-long evolution firmly grounded in graduate study typical of Materials Science and Engineering of a later era. The new department was founded with a graduate level-only curriculum, offering professional training in mining and metallurgy leading to the Degree of Engineer on completion of two years of study. An undergraduate program did not appear in the department until 1923. (Orson Cutler Shepard, '25, a professor and first chairman of the Materials Science Department, and Herbert Hoover, Jr., would be among the first undergraduate students in that new program). The Mining and Metallurgy Department remained as a stand-alone department until 1925, when it was included in the newly formed School of Engineering.

By focusing on the Degree of Engineer as the primary certification for work in industry, Theodore Hoover was backing a long-standing Stanford tradition with that kind of degree. The

"Degree of Mechanical Engineer" and "of Civil Engineer," consisting of an additional year of graduate work and a thesis or report, was outlined in the university's first course bulletin (1891-92). By 1894-95, the "professional degree of Engineer" was permitted for any of the applied sciences, and while more common for mechanical, civil and electrical engineering, included rare awards in chemical engineering (1894, 1907) and metallurgy (1903, awarded to future professor Galen Clevenger). When a Department of Geology and Mining faculty committee recommended formal creation of a degree of Engineer of Mines, following the bachelor's degree, Stanford's Academic Council tabled it without discussion, on the recommendation of Geology chair John Casper Branner. Institution of the degree in 1919, in parallel to the other engineering departments, was something of a coup for Hoover. His major change was to do away with the residency requirement, allowing Mining and Metallurgy students to complete their academic work remotely while employed by mining companies.

The Degree of Engineer was, and still is, a degree offered by a few other universities, including Caltech.

Students with a bachelor's degree in any broad science or engineering program could enroll in this two-year graduate program, leading to the Degree of Engineer in Mining, or Metallurgy, or Petroleum Engineering. Welton J. Crook, featured in a following chapter, would receive his Degree of Engineer (Mining and Metallurgy) in 1919. His was the first Engineer's Degree to be granted by the new department, setting the pattern for graduate study in materials that lasted for more than 30 years.

From 1919 to the late 1940s, virtually all advanced work in Mining and Metallurgy resulted in the Degree of Engineer. About 40 people received that degree during that time period. The list includes many featured in this retrospective, including Welton J. Crook, O. Cutler Shepard, Alden B. Greninger, Servet A. Duran, Alexei P. Maradudin, and Janet Zaph Briggs.

By the early 1950s, with university research more typically funded by federal agencies and foundations, research had become less company-specific and more focused on fundamental scientific problems with broad applications. Such applications required a deeper approach than was envisioned for the Degree of Engineer. This led naturally to the Ph.D. degree becoming the primary advanced degree for the highest-level work. While the Degree of Engineer is still offered by the department, fewer students elected to pursue that degree after 1950.

Similar changes would take place in the undergraduate program. Under the Hoover plan, undergraduates spent their first three years earning 75 units of science (mathematics, chemistry, physics, and geology), 37 units of "broadening courses" (history, foreign languages, English, and Stanford's university-wide required course in "Citizenship"), along with 10 units of electives and 11 units in courses specific to mining and metallurgy. Seniors selected a specialization in mining, metallurgy, or petroleum engineering and took 45 units in relevant coursework. This concept of a broadly educated engineer intent on taking roles in civic leadership and public service has long since faded as a primary goal of engineering undergraduate education.

Faculty from Industry

With the department's creation, its faculty ranks in 1919 were reinforced with three industry-experienced mining engineers, Professors Theodore J. Hoover, '01, his former Stanford classmate, James M. Hyde, '01, and Assistant Professor Carl H. Beal, '12, Engineer, '19, along with the ever-patient Associate Professor Waldemar Dietrich, '13, Engineer, '14. The department had to quickly adapt to the tenuous nature of Hoover's preference for industry-based faculty. Beal left after only one year for a highly successful career in the oil business, as did his successor, John E. (Brick) Elliott, '12.

Frederick G. Tickell (Stanford University Archives)

Stability was provided in part by the arrival of Frederick G. Tickell, a UC Berkeley graduate ('12) with experience in gold mining and petroleum engineering, who joined the department in 1921. Tickell not only remained with the department for decades but served as its second chair from 1936 to 1952, at which point he retired and became emeritus. Unenticed by more lucrative industry jobs, he nevertheless was the department's top fundraiser, attracting major company gifts, bringing in substantial funding for petroleum-related research, and building his own well-equipped laboratory.

While Tickell covered the petroleum division, and Hoover focused on mining engineering and mining economics, Hyde and Dietrich were joined in 1920 by another department mainstay, a young Stanford alumnus with industry experience named Welton Crook, '08, Engineer, '19. Crook, who will be featured in the following chapter, began as a lecturer in 1920, but jumped to associate professor standing the following year.

Dietrich, Crook, and Hyde carried the burden of some thirteen formal lecture classes and eleven lab classes in metallurgy. Crook, for example, taught "Pyrometallurgy of Nickel, Aluminum etc.," "Electrometallurgy," "Metallography," "Pyrometry," and "Metallurgical Calculations." Hyde, an expert on flotation, taught process metallurgy with Dietrich. Together, the three men taught "Crushing Sizing and Classification," "Hyrometallurgy of Gold and Silver," and "Pyro-Metallurgy of Copper, Lead and Zinc," along with "Principles of Metallurgy." Crook and Hyde also taught "Assaying."

James Hyde was not one of Hoover's more productive appointments. Like Hoover, he had been an older undergraduate, and had been a sociable student at Stanford (active in both the Geology and Engineering clubs). He was an instructor in assaying while a senior. After graduating from Stanford, at age 28, he very briefly served on the new mining faculty at the University of Oregon, then spent subsequent years in the mining industry in Mexico and California before joining Minerals Separation, Ltd., in 1910, in its London Office, where he reunited with Hoover, then the company's general manager. Within a year, both had left the company. Hyde returned to the States to pursue mining opportunities in Montana, while Hoover started his London-San Francisco mining and metallurgical consulting offices.

Hyde formed his own mining consulting business, based primarily on a process for extracting zinc from ores by a flotation process that he later claimed to have developed (and which he dubbed the "Hyde Process"). His former employer, Minerals Separation, Ltd, disagreed with his claim, since the process had been developed by numerous engineers working for the British company, and in 1916 successfully challenged Hyde's claim in U.S. courts. Nevertheless, in the United States, Hyde would later be recognized for introducing the flotation process to American mining by the National Mining Hall of Fame and Museum, Leadville, Colorado.

When Hoover returned to Stanford as executive head of the new Mining and Metallurgy Department, he brought the financially troubled Hyde, then working for the California Bureau of Mines, on to the faculty as a full professor. Hyde began his work at Stanford enthusiastically, but soon mixed his teaching with a long-standing interest in local and national public affairs and politics., In 1925, he was an unsuccessful candidate for the Republican nomination for U.S. Senate from California, and subsequently took a year's leave of absence to resume his consulting business. He never returned. Moving to Los Angeles, he held a number of appointed city positions and served on the Los Angeles city council throughout much of the 1930s. (He died in his Hollywood home in 1943).

Hyde would be replaced by future department legend O. Cutler Shepard (see chapter 4). By this time, the department would also begin more extensive use of acting professors and lecturers to bring in temporary assistance from those in industry.

Hoover continued as executive head of the department, as well as engineering dean, until his retirement, as required at age 65, in 1936. In addition to his published work on the *Economics of Mining* (Stanford Press, 1933), he had continued his interest in engineering education, publishing his study with Civil Engineering Department Chair John L. Fish, *The Engineering Profession* (Stanford University Press, 1941), and served as an advisor to Stanford's young Graduate School of Business. Hoover also had become a noted naturalist and early conservationist. He directed a definitive study of the life cycle of the Pacific Coast steelhead trout, in conjunction with the California Fish and Game Commission, and added ornithology and ichthyology to his interests. A portion of his Rancho de Oso, a rare coastal freshwater marsh habitat at the mouth of Waddell Creek that protects rare and endangered species, was

James M. Hyde (National Mining Hall of Fame)

named in his honor as Theodore J. Hoover National Preserve. The remainder of his rancho became part of Big Basin State Park.

Department Names and School Affiliations

Ray Lyman Wilbur, first appointed as executive head of the department of medicine at Stanford, had become Stanford's first "dean," in 1911, with a simple title change. Later, over the course of his first decade in office, President Wilbur indelibly changed the academic organization of the university by forming schools out of clusters of individual programs and departments (Biological Sciences, 1922; Social Sciences, 1924; Engineering, 1925, Business, 1925; Physical Sciences, 1926; Letters, 1926) or by upgrading departments into graduate schools (Medicine, 1911; Education, 1917; Nursing, 1922; and Law, 1924). Stanford was the first major American university to coordinate departments into schools in this manner.

By the early 1920s, Wilbur was considering a school of engineering. The four esteemed and exceptionally influential engineering department heads in 1923—Charles B. Wing of Civil Engineering, William F. Durand of Mechanical Engineering, Harris J. Ryan of

Electrical Engineering, and Theodore J. Hoover of Mining and Metallurgy—all expressed their unanimous disapproval of the establishment of such a school. They insisted that it would lower the standard of the Degree of Engineer, but also that it would dilute the power of departments, and of department heads, and increase "red tape."

Nevertheless, in October of 1924, the president appointed a committee of seven faculty members, chaired by Horacio W. Stebbins, an associate professor of Mechanical Engineering and including Associate Professor Welton J. Crook, to consider creation of a school of engineering. Wilbur took the committee's plan to the Board of Trustees, who approved it on May 15, 1925. Theodore J. Hoover turned from objecting to the school to loyal endorsement, and was named as its first dean. Thus, the School of Engineering, made up of five departments—Civil Engineering, Electrical Engineering, Mechanical Engineering, Engineering Mathematics, and Mining and Metallurgy—was established in 1925.

By 1929, the "and Metallurgy" part of the name of the department was dropped, leading to the name Mining Engineering. This remained the name of the department in the School of Engineering through the academic year 1946-47.

By the early 1940s, a group of alumni of the Department of Geology, called the "Associates"—probably patterned after the successful university-wide alumni fund-raising group, the Stanford Associates—was organized with the objective of influencing the future of the Geology Department. The Geology Associates, along with many Geology Department faculty members, believed that the subjects of mining and metallurgy should return to their close association with geology. They urged creation of a new School of Geology and Mining. World War II brought an end to discussions for the time being, but President Donald B. Tresidder (who had succeeded Wilbur in 1943) moved ahead with the idea in 1946. On November 21, 1946, the trustees approved creation a new School of Mineral Sciences, with the merging of the Department of Geology with the Department of Mining. Thus,

Department of Mining Engineering in 1936. In the circles from left to right are: O. Cutler Shepard, Theodore J. Hoover and Fredrick G. Tickell. (Stanford News Service)

Mining Engineering was removed from the School of Engineering and placed into different "divisions" of Mineral Sciences (separate departments did not exist in the new school). In the first year the study of materials was in a program called Metallurgy, but by 1948-49, the name of the program had been changed to Metallurgical Engineering.

Newly appointed Dean of Engineering Frederick E. Terman was not thrilled with the separation, undertaken before he returned to Stanford from his radar-related war work at Harvard. Terman wanted to retain physical metallurgy in his School of Engineering, by moving the appointments of Welton J. Crook and O. Cutler Shepard (partially) into the Department of Mechanical Engineering. His recommendation was unsuccessful at the time but was the precursor for the move of metallurgical engineering back to the School of Engineering in 1960.

Initially, the program in the School of Mineral Sciences was called Metallurgical Engineering. In 1952, the name was changed back to Metallurgy, but it was changed yet again in 1957, when elevated to the Division of Metallurgical Engineering. Just one year later, in 1958, a new Department of Metallurgical Engineering was created within the School of Mineral Sciences, with O. Cutler Shepard as its executive head.

Finally, in 1960, when the Board of Trustees approved its transfer back to the School of Engineering, it acquired a completely new name: Materials Science. (Shepard retained his position as executive head.) That name continued until 1971, when it was changed to the Department of Materials Science and Engineering, a name it has retained for nearly 50 years. Below is a table outlining the tortuous organizational and naming history of the department from 1919 to the present.

From Mining and Metallurgy to MSE: Changing Names and Schools

Years	Department	Programs/Divisions	School
1919-1924	Mining and Metallurgy	-	Independent
1925-1928	Mining and Metallurgy	-	Engineering
1929-1946	Mining Engineering	-	Engineering
1947-1957	--	Metallurgy/ Metallurgical Engineering	Mineral Sciences
1958-1959	Metallurgical Engineering	-	Mineral Sciences
1960-1970	Materials Science	-	Engineering
1971-present	Materials Science & Engineering (MSE)	-	Engineering

The Thomas F. Peterson Laboratory, 2000. (MSE Dept.)

The Mining and Metallurgy Building

When the Department of Mining and Metallurgy was founded in 1919, it remained in the Mining and Metallurgy Building, a long sandstone building that had been built in stages near the corner of Lomita and Panama streets, behind the Quadrangle near Geology corner. The construction of the first wing of the building was completed in 1900 and was originally used by Civil Engineering. The second wing, called the Metallurgy Laboratory, was added in 1908, after considerable damage had been done to the building by the 1906 earthquake. This addition made an L shaped building. Another wing was added in 1915, making a U-shaped building. By this time, it was called the Mining and Metallurgy building.

The space in between the two arms of the U was used for faculty parking through the 1930s and was still called "The Courtyard" in the 1960s and later. In 1949, with sponsored research creating the need for more laboratories, the courtyard was covered and made into laboratory space. That space was upgraded several times in the 1950s and 1960s to improve the laboratory areas there. It had been one open laboratory when Robert A. Huggins arrived on the faculty in 1954, and Huggins organized and led the building of separate laboratories in the space, an effort that signaled the establishment of central laboratory facilities in the decades ahead.

The L shaped part of the building was renovated in 1963 and was named the Peterson Laboratory, after Thomas F. Peterson, whose family provided funds for the renovation. The wing that had been added in 1915 was not renovated, and by the 1980s that space was nearly useless, except for a room used for the ping-pong table and parties. That limitation was rectified in 1990 when the 1915 wing of the building was also renovated and made suitable for offices and light laboratories. The very end of that renovated wing was used by the Center for Integrated Facility Engineering (CIFE), which helped to secure funding for the renovation.

Department offices and some faculty offices and laboratories continued to be housed in the Peterson Laboratory until 2008, when those offices and labs were moved into the department's present home, the William F. Durand Building. (The Durand Building, first called the Space Engineering Building when it was completed in 1969, initially housed interdisciplinary work in aeronautics fostered by Professor William Durand, but now houses a variety of engineering units.)

The Peterson Laboratory was later renovated and today is occupied by the "d.school," the Hasso Plattner Institute of Design at Stanford.

Part II ~ From Metallurgy to Materials Science

Chapter 3

Welton J. Crook: Steel Metallurgist, Par Excellence

The story of materials science would not be complete without a report on Welton J. Crook, a colorful individual who would personify metallurgy at Stanford for decades and have a significant impact on the steel industry with which he was closely associated. His career at Stanford extended from his undergraduate studies of metallurgy in the Geology and Mining Department, through his faculty service in the new Department of Mining and Metallurgy. Crook taught steel metallurgy and conducted research on steel well into his retirement years, during which time he maintained an active consulting business using emission spectroscopy for the chemical analysis of steels.

Student Years and Industrial Career

Welton Crook arrived from his native Australia in 1908 as an undergraduate student in mining. He had been born in Melbourne, in 1886, where his father was an architect and engineer who had worked with Thomas Welton Stanford, younger brother of U.S. Senator and University co-founder Leland Stanford. Crook was named after Welton Stanford, who was also his godfather. Surely that influence brought him to study, at the age of 22, to Stanford University, after his training at the Melbourne School of Mines and his four years' service in the Australian army.

In his review of the history of metallurgy at Stanford, written in 1968, Crook wrote about his first days at Stanford with another "Aussie," Henry Lyne, who also was studying mining. Many of the Geology and Mining Department's students and faculty spent their summers working in silver mines in Mexico. To keep their Spanish sharp, Professors Galen Clevenger and Luther Bahney, both fluent in Spanish, developed a practice of giving their mining and metallurgy lectures in Spanish as well. Crook recalled that students spoke and swore in Spanish in the laboratories, including rather rude nicknames for each other. Many expected to get jobs at Mexican mines after graduating. To accommodate Crook and Lyne, however, lectures were eventually, and thankfully, given in English.

Mining and metallurgy were closely intermingled in the department, with an emphasis on mineral and petroleum extraction, Crook later recalled, but distinct from geology (which also included paleontology). Assaying, once firmly embedded in the Chemistry Department, was gradually moving to the mining and metallurgy program as its own sub-rosa lab course, taught by lab assistant Crook,

The Cosmopolitan Club, Stanford University, 1907. President David Starr Jordan sits in the center, with Welton J. Crook on the right. (Stanford University Archives)

grew in popularity. In time, even the assaying equipment was moved over to the metallurgists, "where it belonged," said Crook.

Crook, with Knight Starr Jordan, son of Stanford president David Starr Jordan, graduated with an A.B. degree in mining and metallurgy in 1911, the only students to receive that particular degree. Crook had refused to accept a degree in geology and mining on the grounds that he had not completed sufficient work in geology, a concession that department head John Casper Branner granted reluctantly.

After graduation, Crook took a position as chief chemist and metallurgist for the Pacific Coast Steel Company, at the new company's steel works in San Francisco. That year he also married Maud L'Ampheres, a fellow geology student, and an assistant in Stanford's Department of Hygiene (precursor to today's Department of Athletics). Shortly thereafter, in 1913, on the recommendation of Professor Clevenger, he was made professor of metallurgy at the South Dakota School of Mines, and head of the Metallurgy Department and Director of the South Dakota State Metallurgy Experiment Station. The youthful-looking Crook, then 27, grew a moustache to look older. (As an aside, Crook's arrival at the South Dakota School of Mines was just two years after the famous metallurgist Zay Jeffries had graduated from that school.)

Sensing little future at the South Dakota School of Mines, in 1917 Crook returned to his position with the Pacific Coast Steel Company, where he was asked to improve their process of manufacturing much-needed artillery shells for the French army. During these years in San Francisco, he completed graduate work at Stanford, receiving its degree of Engineer in 1919. His Engineer's thesis, entitled "A Study of the Hot-dip Galvanizing Process," is the first on the list of metallurgical theses at Stanford.

Just one year later, in 1920, Crook joined Stanford's recently established Mining and Metallurgy Department as an associate professor, replacing Clevenger. Crook's appointment was a sign of changing times in the department. Where his predecessors had worked with gold and silver (Clevenger) and copper, lead and zinc (Bahney), Crook's work with Pacific Coast Steel had pointed him in another direction: "Steel was the new thing." He would be promoted to professor in 1927.

Faculty Career: Teaching, Research, and Collaborations with Industry

In his anecdotal history of metallurgy at Stanford, Crook tells his story of getting started as a new faculty member at Stanford. According to Crook's account, there was practically no usable equipment in the department for his type of work, and there were only a couple of very old, low-power bench microscopes of the type that Albert Sauveur had used decades before at Harvard. Department head Ted Hoover had, inexplicably, sold or disposed of any furnaces suitable for heat treatment.

A few years later, seeking to rectify the situation, Crook went, hat-in-hand so to speak, to Judge E. M. Wilson, president and principal owner of Pacific Coast Steel Company, the company that he had deserted to take the Stanford job. Crook told Wilson of his dilemma and asked for help. After reminding the young professor that he had urged him not to accept the Stanford job and that he should have stayed in the steel business where he belonged, Wilson took out his checkbook and wrote a check for $5,000 (the equivalent of $75,000 in 2019 dollars). The check might help a little, Wilson said, but he still thought Crook had been foolish to get himself into this situation. Such was the nature of faculty start-up packages in the 1920s.

Faculty and students in Mining Engineering in 1938.
In the circles, from left to right, are O. Cutler Shepard, Welton J. Crook, and Frederick J. Tickell.
(Stanford University Archives)

Faculty in the Mining Engineering Department in the early 1940s. From Left to right are: O. Cutler Shepard, Welton J. Crook, Fred Humphrey (Mining Engineering), Kenneth Schellinger (Extractive Metallurgy) and Fredrick G. Tickell (Petroleum Engineering). (Stanford News Service)

With funds in hand, Crook happily bought a Leitz metallograph, hardness testers, and heat treatment furnaces. Throughout his career, Crook would value a well-outfitted laboratory.

A peek at the *Annual President's Reports* of those early years reveals other gifts to the new department – fellowships in mining were set up by Lou Henry Hoover, '98, (Mrs. Herbert Hoover), a former star geology pupil of J. C. Branner, and by Theodore Hoover, and funds were also obtained to support field research for the petroleum industry – but Wilson's very substantive gift was the first to support an early materials program, and played an important role in updating the department's laboratory.

Crook had returned to Stanford as an expert in steel metallurgy, and he continued in that line of research for the rest of his career. Little is known about his early metallurgical work, but a 1972 letter about Crook by John E. (Brick) Elliott, one of Crook's classmates at Stanford, gives a glimpse of Crook's activities. Elliott was a petroleum geologist and had invented a new type of drill bit and other oil drilling apparatus. After several years as chief metallurgist for Shell Oil of California, in the early 1920s he had formed the Elliott Core Drilling Company to exploit that invention. Crook was a consultant for his company, developing an inexpensive but effective hard-facing metal for oil well drill bits. The hard-facing metal was probably a heat-treated alloy steel with a high carbon content.

Crook's work for Elliott was an effort to avoid the high costs of a new "secret" hard-facing metal that had been invented in Germany. It is almost certain that the new metal to which Elliott referred in his letter was an alloy consisting of tungsten carbide particles in a cobalt matrix. Cemented carbides had been invented at the Osram Lamp Works in 1923 and were marketed as hard-facing materials by Krupp AG in 1927 under the trade name "Widia." These alloys were later exploited by Carboloy Inc. (a subsidiary of General Electric Company), which was led by the famous metallurgist, Zay Jeffries. Crook's work on steels in the 1920s resulted in a U.S. patent for a steel to be used for "sucker rods," the metal rods that connect the surface and down-hole components of a reciprocating pump in oil wells.

Later in the 1920s, Crook published papers on "Alloy Steels and their Uses" (*Journal of Chemical Education,* 1927) and "Heat Treatment of Low Carbon Steel" (*Metal Progress,* 1930) that reflected his continuing interest in education about steels and their uses. It must have been during that time that he acquired from his industrial collaborators a complete set of commercial alloy steel bars for use in teaching and research in the department. The bar form of these samples made them suitable for end-quench hardenability studies that were common in laboratory courses of the time. That set of steel bars remained in one of the stockrooms of the Mining and Metallurgy building for most of the rest of the twentieth century.

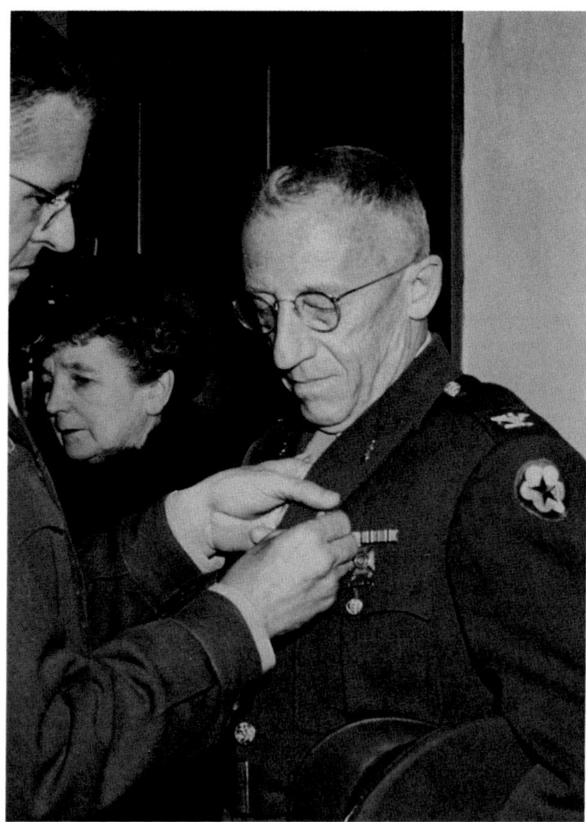

Colonel Welton J. Crook receiving the Army's Legion of Merit. (Stanford News Service)

Crook also acquired large photographic displays of optical micrographs showing different microstructures that could be created in different steels with different kinds of heat treatment. These were used to show students the kinds of microstructures they should find in their lab experiments. They are still in use as art objects in different offices on campus.

By the mid-1930s, Crook had served on the faculty for 15 years and was in line for a sabbatical. He spent it at the Metallurgical Research Laboratory at the University of Romania in Bucharest, Romania. Already accomplished in the science and technology of steel, he earned during his year at the University of Romania the degree of Doctor-Engineer for his work on the constitution of slags in steel making. He held the distinction of receiving, in 1936, the first doctorate in engineering awarded by that university. His dissertation, entitled "Experimental Research on the Mineralogical Constitution and Chemical Reaction of Slag in the Making of Steel," was published in several journals in the late 1930s, three of them closely associated with École Polytechnique and one the French journal *Comptes Rendus.*

After World War II, Professor Crook preferred the title of "colonel," acquired through his service in the U.S. Army. He had joined the U.S. Army Reserves (Ordnance) as a major in 1927, and later was promoted to lieutenant colonel. During the war, he was promoted to full colonel when he was appointed a technical officer in the Army's Engineering and Steel Casting Division at the Rock Island Arsenal in Illinois. While working there in 1942, Crook applied for a U.S. patent (received in 1944) for one of his developments on the heat treatment of steel. For his war work, he was given the

Army's Legion of Merit, its fourth highest award. The colonel was said to have kept a military revolver in his desk drawer to display to selected visitors.

Retirement Years

In the latter part of his career, Crook turned more of his attention to consulting, maintaining an emission spectrograph in his lab for detecting trace elements in steels deleterious to mechanical properties. (When a steel sample is heated to a very high temperature with a carbon arc, the light emitted from the glowing metal, after passing through an appropriate prism, forms a spectrum that reflects the presence of the constituent atoms.) Crook provided much needed services to local steel companies through his work using emission spectroscopy.

Colonel Crook served on the Stanford faculty until his retirement in 1951. On a subsequent trip to Australia, with a stop in Hong Kong, he became intrigued with the abacus. On discovering that there were no books in English describing the workings of the device, over a two-week period he wrote a booklet entitled *Abacus Arithmetic*. This booklet, published by Pacific Books in 1958, proved to be an instant best seller, with more than 80,000 copies sold.

By the 1960s, Crook's lab was located in the attic of the still un-renovated section of the Mining and Metallurgy building. He also made use of chemical laboratories in the main part of the building. In doing so, he maintained his own clean beakers and test tubes, which he guarded like a mother hen, as Alan Ardell, a graduate student at the time, would discover. Ardell had the temerity to ask the aging Colonel if he might use one of Crook's clean beakers rather than find one of his own. Crook's response was not recorded, but it was probably not fit to print anyway. Ardell looked for another beaker.

Colonel Crook died in 1976 at the age of 90. He was honored by the department in 1960 with the establishment of the Welton J. Crook award, which goes each year to the outstanding senior graduating from the Materials Science and Engineering Department.

Col. Welton J. Crook, at his spectrograph, in 1967. (Stanford News Service)

Chapter 4

O. Cutler Shepard: From Assayer of Gold and Silver to Head of the Materials Science Department

The arc of progress that led from metallurgy in the 1920s to materials science in the 1960s can be told by tracing the career of Orson Cutler Shepard. Shepard personified that era at Stanford, from having been one of the first to enroll in the undergraduate program in the Mining and Metallurgy Department in 1923 through his 40 years on the faculty (1928-68). Along the way, Shepard initiated research with graduate students, trained the first doctoral students in the department, and started the department's first federally sponsored research projects. He held the metallurgy program together in the difficult 1950s, becoming the first executive head of the reorganized Department of Materials Science in 1961.

O. Cutler Shepard in his office, 1946. (Stanford News Service)

Early Years and Student Days

Cutler Shepard was born in 1902, in the small San Joaquin Valley farming community in Del Rey, California, near Fresno. Three years later, his family moved to San Jose, where his grandfather had a fruit farm at the corner of McLaughlin Avenue and Story Road (just a few blocks from where the author of this retrospective lived 50 years later). Shepard remembered being at his grandfather's farm on April 18, 1906, when the great earthquake hit the Santa Clara Valley. His father, Orson Earnest Shepard, was an early auto dealer who complained that people just wanted to test drive the cars, not buy them. After his grandfather died, in 1910, Shepard and his family moved to the farm on McLaughlin Avenue, where they grew prunes, pears, and walnuts.

Shepard attended the two-room McKinley school before enrolling, in 1915, at San Jose High School on Washington Square in downtown San Jose, where the San Jose Normal School (today's San Jose State University) also was located. He graduated from San Jose High School in 1919. Shepard started college at the College of the Pacific (later the University of the Pacific), which was then located in San Jose. He later recalled

Cutler Shepard (far left), with Alden Greninger (next to Shepard) and other students at the back of the Mining and Metallurgy building in about 1930. (Stanford News Service)

that he was able to commute easily from school to home to help with work on his parent's farm. In 1921, he enrolled at Stanford, but subsequently missed a couple of quarters because he had to help at the family farm. A year later, his family rented out the San Jose farm and moved to Palo Alto, so that Shepard could focus on his studies at neighboring Stanford.

Shepard graduated from Stanford in 1925 with an A.B. in Mining and Metallurgy. During the spring quarter of his senior year, he had applied for jobs in industry, aiming to acquire metallurgical experience. While in the Sierra foothills town of Jackson, California, doing some work for a spring quarter course entitled "Mine Surveying," he received a telegram from Anaconda Copper Mining Company in Montana, offering him a job in their research department. The catch was that the offer was good only if he could come to work immediately Shepard worked non-stop, around the clock, so that he could leave for Montana the following day.

Shepard worked at Anaconda for about a year before moving on to a position as an assayer and assistant mill superintendent at the San Luis gold and silver mine in Mexico. The mine was deep in the mountains, a two-day trip by horseback from the nearest road. Shepard conducted tests and made calculations pertaining to the construction of a new plant there. After a year and a half, he returned to visit his family in Palo Alto, and while there, he was encouraged by Theodore Hoover to come back to Stanford for graduate work. After returning to Mexico for several months to complete his job, he returned to Stanford for graduate study, finding time to marry Grace Newland in 1927. He received the degree of Engineer in Metallurgy in 1928, with an Engineer's thesis entitled "A Hydro-metallurgical Process for the Treatment of Chalcopyrite Concentrates from the Walker Mine." After a summer job at the Walker Copper Mine in Plumas County, California, he accepted a position as instructor in the Mining and Metallurgy Department at Stanford. That was the start of his 40-year career with the department.

Teaching and Research with Students

By 1931, Shepard had advanced to the rank of assistant professor and he was starting to attract graduate students in the field of physical metallurgy. That year, one of his students, Alden B. Greninger, completed an Engineer's thesis entitled "An Investigation of the Electrodeposition

Mining Engineering graduate students pose with Bailey Willis (left circle) and Cutler Shepard (right circle) in the late 1930s. (Stanford News Service)

of Chromium and the Structure of Chromium Deposits." Greninger went on to Harvard for his doctoral work on the crystallographic aspects of phase transformations in copper-zinc (CuZn) alloys. At Harvard, Greninger devised a convenient method for determining the angular relationships between crystallographic planes from an x-ray diffraction photograph, later called the Greninger chart. After graduating from Harvard in 1935 and spending a few years there on the teaching faculty, Greninger joined the General Electric Company (GE) in 1941 and spent much of his career at GE working on nuclear power. He also served two years with the Army's Manhattan Project during the Second World War. In February of 1954, Greninger was the featured speaker for the dedication of the first engineering building at San Jose State College, where the author of this retrospective was to enroll the following September.

In 1936, Theodore J. Hoover retired, both as Dean of Engineering and as Executive Head of what was now called simply the Mining Engineering department. Samuel Morse replaced Hoover as Dean and Fredrick George Tickell, a specialist in petroleum engineering, became department head. Welton J. Crook remained on the faculty with a focus on iron and steel, while Shepard focused on metallurgy more generally. Shepard continued his work with graduate students throughout the 1930s, directing M.S. and Engineer's theses on physical metallurgy topics.

As an indication of the central role that mining and metal processing continued to play in the department in the 1930s, one need only consider one of the first textbooks to come out of the department. In 1940, Shepard and metallurgist and former department faculty member Waldemar Fenn Dietrich published the book *Fire Assaying*, with the McGraw-Hill Book Company. The book described various methods used to determine the compositions of certain metal alloys, especially the noble metals like gold, silver, and platinum. Dietrich, at Stanford from 1915 to 1930, was then teaching mining and metallurgy at Sacramento Junior College. Shepard, now an associate professor at Stanford, also was a visiting associate professor at Massachusetts Institute of Technology (MIT), where he had gone for a sabbatical leave. Interestingly, the committee that approved publication of Shepard and Dietrich's book as a metallurgical text included Stanford alumnus and former faculty member Dorsey A. Lyon, as well as the famous metallurgists Bradley Stoughton (after whom the ASM Young Teacher award is named) and Zay Jeffries, mentioned earlier.

Shepard's sabbatical leave at MIT (1940-41) came at a time when important changes were taking place in the Metallurgy Department there. Under new department head John Chipman, MIT metallurgy courses relating to plant operation were being replaced with new courses on the application of physics and chemistry to metallurgy. Shepard participated in seminars in physical metallurgy with faculty from MIT and Harvard, and was able to reconnect with Alden B. Greninger, still at Harvard. Shepard also developed an interest in x-ray diffraction and other modern characterization techniques that he would bring back to Stanford.

Ladislas L. (Bill) Marton of the Electrical Engineering Department with his electron microscope. (Stanford University Archives)

Returning to Stanford in 1942, Shepard initiated significant changes in the metallurgy curriculum that reflected his MIT experience. He created the course Met.E 104, "Introduction to Physical Metallurgy" which became the service course in materials for engineers for nearly 20 years. This course, and other graduate courses in metallurgy, reflected the transformation taking place more broadly in the field. Academic work on specific industrial problems or plant operations moved toward research on more basic science that would become the basis for creating future industries. This transformation was echoed by the move away, later in the decade, from the degree of Engineer to the Ph.D.

Shepard also offered a course entitled "Physics of Metals," which was an updated version of Dorsey Lyon's original course, given almost 40 years before. Shepard also collaborated with Ladislas L. (Bill) Marton, an associate professor of electron optics in the Electrical Engineering Department, who was working on the design and construction of an electron microscope. Together they offered a course entitled "X-rays and Related Methods in Metallurgy."

The Hungarian-born Bill Marton was an early pioneer in transmission electron microscopy. He had immigrated to the United States by way of Belgium at the outset of World War II. He initially worked on electron microscopy at the Radio Corporation of America (RCA), where he designed and built a two-stage microscope. His design was more difficult to produce than others being considered, so RCA did not pursue his project. Marton moved to Stanford, where, to support this work, a Division of Electron Optics was formed in the Department of Electrical Engineering. Marton received a notable grant of $65,000 from the Rockefeller Foundation to design and construct an improved electron microscope, and he was one of the first to develop a wet cell for the electron microscope and to image bacteria with the apparatus. His complex microscopes, however, with their multiple lenses, were considered too far ahead of their time for commercial production.

Shepard's First Ph.D. Students

The 1940s saw the first Ph.D. degrees granted to Stanford students for work in the field of materials. O. Cutler Shepard's first two doctoral students were both from Turkey. Nevzat Tiner received his Ph.D. in 1945 with a dissertation entitled "Studies on Superheating of Magnesium Alloys," while Mehmet Hayri Erten received his degree in 1946 for a dissertation entitled "Effect of Shot Peening on the Fatigue of Turbine Blade Materials."

Mehmet Hayri Erten, photographed during his time at Stanford. (M. H. Erten)

Mehmet Erten was born on September 15, 1920, in Antep (today's Gaziantep), Turkey. After studying at the University of Minnesota, he came to Stanford for graduate work, receiving his M.S. and Ph.D. in 1945 and 1946, respectively. He worked in Turkey for several decades before becoming a professor at the University of Missouri at Rolla. He retired in 1990 and was living in Virginia at the time of this writing.

Little is known at this time about Tiner's biography or later career.

The works by Tiner and Erten set in motion an academic activity at Stanford that has now produced more than 2,300 M.S. and Ph.D. degrees in materials since 1945. The graph below shows the cumulative count of M.S. and Ph.D. degrees granted by the department since the mid-1940s. In the last few years, about 40 M.S. and 22 Ph.D. degrees have been granted each year, a little higher output than the average since 1960. The output of advanced degrees from the department has been relatively constant since the late 1960s, even though there have been fluctuations in the size of the MSE faculty (from a traditional faculty head-count of 12 in the 1960s

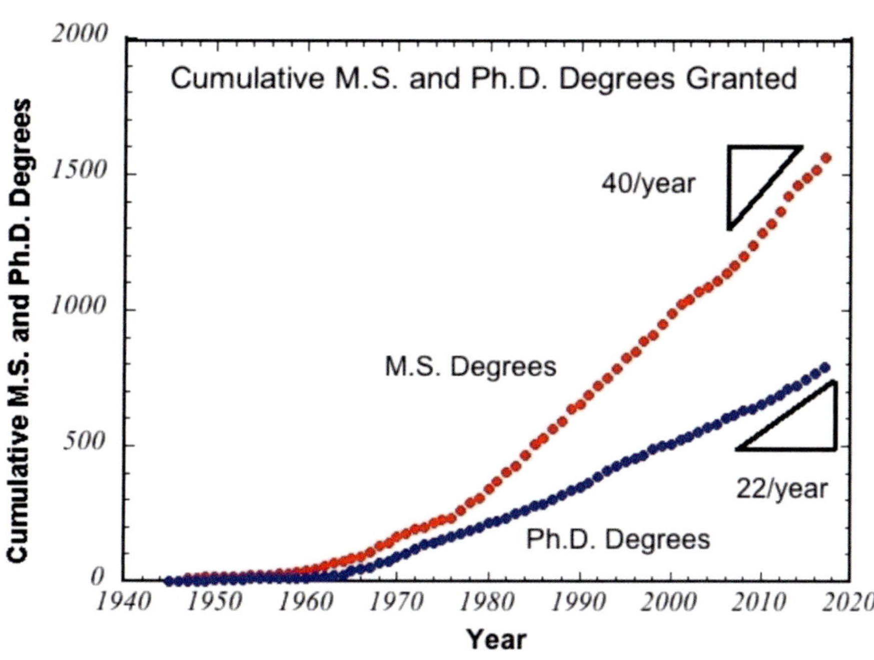

and 1970s, to a low of 6 for a short period in the 1990s, to a high of 17 in 2018). A number of factors explain why the degree output and faculty headcount do not track more perfectly. One factor is that, since the 1960s, a significant number of MSE doctoral students have been advised by faculty outside the MSE department. At the same time, more and more MSE faculty have directed the work of doctoral students in other academic departments, especially since the 1980s.

Early Post-War Sponsored Research

Although Galen Clevenger had had a sponsored research project in the teens, federally sponsored research in materials, now an important source of funding for graduate students, did not get started until after the Second World War.

Ralph Parkman, one of Shepard's early students (about whom more is written below) reported to the author in 2007 on those early days of sponsored research after the war. Parkman wrote:

> Around 1946, as I recall, Shepard had gotten a small NSF grant for a creep project in which the specimens were surrounded by a controlled atmosphere. This was not too common and the work seemed to have gotten the attention of certain government agencies, including some connected to Admiral Rickover, who was involved in a crash project to design and build the first nuclear submarine. There were many contractors including General Electric (GE). Representatives from GE came out to talk to Shepard and his students and to talk to people from the Berkeley Metallurgy Department. The result was that Stanford and Berkeley (using different designs) were to test a variety of metals under stress, and in contact with liquid metals in a controlled atmosphere, such as might be used in future high-temperature heat transfer systems.
>
> That project, which involved a great deal of money for the times, started in July 1948. Our design (mainly Tom Paine's ideas) proved superior to those of Berkeley. The project kept going in many directions under successive sponsorships of the Office of Naval Research (ONR) and the Atomic Energy Commission (AEC). There was enough money for a whole new lab for the apparatus, and to support many students, including the eight Ph.D.s mentioned in Shepard's historical piece. (Parkman, 2007)

As noted in Parkman's letter, Thomas O. Paine (Tom Paine) was a key figure in getting Shepard's sponsored research project started. Shepard's work involved an effort to understand and control the effect of environment on the creep of metals. He discovered that extremely pure argon was still not sufficiently pure to avoid atmospheric degradation of some metals. As a remedy, he suggested using molten lead, surrounded by argon gas, to protect the test specimens. He discovered that excellent heat conduction in the molten lead produced more even temperatures in the test pieces, and that more even temperatures led to much greater uniform elongations, even for plain carbon steels. As already seen in Parkman's letter, Paine's work led to a series of contracts from ONR and AEC that supported several early Ph.D. students, including Parkman.

Shepard's Doctoral Students of the 1950s

Shepard's work with doctoral students took off in the late 1940s when he started to secure contracts from various federal agencies for fundamental research related to the development of nuclear power and other kinds of power plants.

Although Shepard did not have a doctoral degree himself, in less than a decade he trained nearly 10 Ph.Ds. who would play important roles in shaping the field of metallurgy and materials in industrial, governmental and academic institutions.

The most prominent of Shepard's students would be Thomas O. Paine, who eventually would become the Administrator of the National Aeronautics and Space Administration (NASA). Paine was born in Berkeley, California, in 1921. He graduated from Brown University in 1942 with an A.B. degree in Engineering. Serving as a submarine officer and qualified deep-sea diver in the Pacific during the Second

Thomas O. Paine (NASA)

World War and the later occupation of Japan, he was awarded the Commendation Medal and Submarine Combat Insignia with stars. In 1946, he headed to Stanford to study physical metallurgy, receiving his M.S. degree in 1947, and Ph.D. in 1949. Paine subsequently joined the GE Research Laboratory in Schenectady, New York, where he worked on magnetic and composite materials. For the next two decades, he took on positions of increasing responsibility with GE, first in New York and later in California. Nationally recognized in both industrial and professional circles, Paine was appointed Deputy Administrator of NASA in 1968, and was confirmed as its third administrator in March 1969, in time to oversee the first moon landing the following July. During his brief leadership of NASA, the first seven Apollo manned missions were flown, involving 14 astronauts who traveled to the moon and four who walked on its surface. Paine resigned from NASA in 1970, returning to GE as vice president and group executive of the Power Generation Group, where he remained until his retirement in 1976. He died at his home in Los Angeles in 1992.

The sponsored research projects that Shepard started in the late 1940s provided support for a growing number of graduate students in metallurgy, many of whom, including Tom Paine, Ralph Parkman, and Jack Lewis, were veterans returning from World War II. Ralph Parkman (about whom more is written below) had followed up Paine's work by focusing on the attack on metals by liquid lead and bismuth. He finished his Ph.D. in 1952 and taught for three years at Stanford before joining the faculty at San Jose State College. Jack Lewis finished his degree in 1951 and worked at North American Aviation for many years, becoming an expert on titanium alloys. Others who worked on Shepard's projects during that time were Kenneth L. Keating, who became a professor at the University of Arizona; Edward P. French, who worked in gas dynamics in southern California; and Steven J. Rothman, who worked at the Argonne National Laboratory near Chicago and became the editor of the *Journal of Applied Physics* later in his career. The other Shepard Ph.D. students during that period included Willis L. Schalliol, who later owned a metal casting company in the mid-west; and Donald R. Mash, who worked at the Lawrence Livermore Laboratory before becoming the Dean of Engineering at the University of Portland and, later, Professor at the California Polytechnic School in San Luis Obispo.

Another student from the 1950s, Alexei A. Maradudin, deserves special mention. His father, Alexei P. Maradudin, a Russian immigrant, had studied metallurgy at Stanford in the 1920s and had gone on to a successful career as a welding metallurgist for Standard Oil of California. The son, Alexei A. Maradudin, received his B.S. and M.S. degrees in physical metallurgy from the department in 1953 and 1954, respectively. He then moved to the Department of Physics at Bristol University under a Marshall Scholarship, with a plan to work on dislocation theory. He received his

Ph.D. from Bristol in 1957 after working with the famous physicist Charles Frank. A.A.. Maradudin went on to a brilliant academic career, mostly at the University of California, Irvine. Ralph Parkman deserves special mention in this retrospective because of his role in furthering the study of metallurgical engineering at Stanford and for establishing metallurgical engineering, later called materials engineering, at San Jose State College (today's San Jose State University).

Ralph Parkman, as a graduate student in 1952. (R. Parkman)

Perhaps it was predetermined that Parkman would become a metallurgist: he was born in Penzance, Cornwall, England, a county that had been known for mining since the Bronze age. (He in fact came from a family of fisherman.) When he was four years old, Parkman came with his family to the United States, settling in Erie, Pennsylvania. In the late 1930s, he studied engineering at the Erie Center, a satellite campus of the University of Pittsburgh, and later transferred to the main campus in Pittsburgh where he received his B.S. degree in Metallurgical Engineering in 1941. After working as a metallurgist at the Crucible Steel Company of America for the first few years of the war, he joined the U.S. Navy and served in the Philippines. Following the war, he briefly returned to industry before coming to Stanford as a graduate student in 1947, where he worked with O. Cutler Shepard, Tom Paine, and others on the behavior of structural metals in liquid metal environments. He received his M.S. degree in 1949 for work on protective metallic

Garden party for students and colleagues at the home of Professor O. Cutler Shepard in about 1951. In the first row, from left to right: O. Cutler Shepard and Dolf Hubbard (M.S. 1951). Kneeling between the rows is Jack Lewis (Ph.D., 1951). Standing, from left to right: Cynthia Shepard (daughter), Ralph Parkman (Ph.D., 1952), and Lt. Col. Chuck Beaudry (M.S. 1951). (Stanford News Service)

coatings on structural alloys and completed his Ph.D. in Metallurgical Engineering in 1952, with a dissertation entitled "The Attack of Solid Metals by Liquid Lead and Liquid Bismuth."

After teaching in Stanford's Metallurgy Program for three years, Parkman joined the faculty at San Jose State College where he co-founded the Metallurgical Engineering Program. That program later evolved into its Materials Engineering Department (precursor of the current Department of Biomedical, Chemical and Materials Engineering). Parkman served on the faculty at San Jose State for more than 30 years, including some years as department chairman in the 1960s, before becoming professor emeritus in 1985. In the latter part of his career, he also was a co-developer of an interdisciplinary course entitled "Cybernation and Man," which examined the effects of the emerging information technologies on society. This led to his 1972 textbook, *The Cybernetic Society* (Elsevier).

In addition to his contributions to materials teaching and research at Stanford, Ralph Parkman contributed enormously to Stanford's MSE department by preparing a number of San Jose State undergraduate students who would later successfully complete their own Ph.D. degrees at Stanford. These include:

William D. Nix, 1963

John L. Brimhall, 1964

W. Robert Johnson, 1969

Patrick P.P. Pizzo, 1975

Bruce E. Leibert, 1977

Michael J. Mills, 1985

Darcy A. Hughes, 1986

Search for a New Direction in the 1950s

The 1950s proved to be a difficult time for the study of metallurgy and metallurgical engineering at Stanford, as the department experienced four different reorganizations in just that one decade. A review of *Courses and Degrees* shows only two regular faculty members covering physical metallurgy during most of that time. After Welton J. Crook retired in 1951, only Professor O. Cutler Shepard and Assistant Professor Lewis Dana Hall remained.

Hall taught courses on the physics of metals and reaction kinetics in solids, and he was the first to teach dislocation theory at Stanford. Kenneth L. Keating and Ralph Parkman served as instructors for the period 1952-1955. After serving two years, however, Hall left to work for Varian Associates. He was replaced in 1954 by Robert A. Huggins, whose contributions are described in the following chapter. Shepard and Huggins continued as the faculty of two for the rest of the decade.

From 1955 to 1957, A. Kenneth Schellinger was listed as an associate professor in the metallurgy program. His focus was on extractive metallurgy and mineral processing, and he was listed in both the Metallurgical Engineering and Mineral Engineering Divisions of the School of Mineral Sciences, so he was not in a position to help much with the parts of metallurgy that would lead to materials science.

While Shepard was to serve on the faculty for eight more years, as professor and executive head of the department, the direction of the department, and the beginning of the era of materials science at Stanford, would be defined by Robert Huggins and, after 1958, by newly appointed faculty. Shepard's duties as executive head limited his own research activities, so he later collaborated with Oleg Sherby and Sherby's students on the use of torsion to assess the ductility of steels at elevated temperatures. By 1967, several senior faculty members had joined the department, and Shepard might have felt that the building of the Materials Science Department was essentially complete. With that phase of his work finished, Shepard stepped down as chair and prepared for retirement.

Shepard's Retirement Years

Shepard retired from the active faculty in 1968, an occasion that was celebrated by the department with a memorable symposium and dinner in his honor. The symposium included a talk by the author of this piece on "The Evolution of Metallurgy at Stanford;" one by Huggins on "The Evolution of Materials Science at Stanford;" and one by William A. Tiller, the new head of the department, on "The Future of Materials Science at Stanford." (Tiller had been appointed to the faculty in 1964.)

Engineering Dean Joseph M. Pettit's brief remarks at the dinner were followed by what turned out to be a long address by Shepard's colleague, Col. Welton J. Crook. Crook had written an impressive review of metallurgy at Stanford for the occasion entitled, "A Partial History of Metallurgy at Stanford University." His paper is an important historical document, but the reading of the entire 19-page paper as an after-dinner speech to assembled guests who were well fed and probably sleepy was not one of Crook's better ideas. There was much shuffling of feet as the organizers agonized about how to terminate the proceedings. Having started his Shepard speech with the seminal events of 1891, Crook mercifully stopped when he got to about 1940 and said, "Here endeth the first lesson."

Shepard lived for nearly 30 years after becoming professor emeritus in 1968, continuing to come to the department for seminars and to offer advice when asked. In 1979, he provided the department with endowment funds to establish an award to be given each year to the most outstanding recipient of the Master of Science degree or degree of Engineer in Materials Science and Engineering. In recognition of that generosity, and of Shepard's life-long contributions to the department, the award was named in his honor: The O. Cutler Shepard Award. With that award, and with the Welton J. Crook Award for undergraduates, the contributions of the two people who guided the evolution of the department from Mining and Metallurgy to Materials Science, were recognized in a permanent and fitting way.

Col. Welton J. Crook and O. Cutler Shepard, 1968. (Stanford News Service)

In December 1992, on the occasion of Shepard's 90th birthday, the department again celebrated his life-long contributions to metallurgy and materials science at Stanford with a symposium and a banquet in his honor. The symposium included talks on the history of metallurgy and materials science by the author of this piece; Shepard's role in teaching and research by Oleg Sherby; reflections by a former student, M. Lea Rudee, dean of engineering at U.C. San Diego at the time; and comments on the growth and development of the department in the 1960s and 1970s by William A. Tiller. Shepard, along with his wife Grace and their children and grandchildren, were honored at the banquet, which was a most enjoyable social occasion.

In preparation for this special birthday celebration, Shepard wrote his own history of the evolution of the department, which was published as a brochure entitled, *Historical Sketch of the Department of Metallurgy and Materials Science,* and distributed to attendees at his birthday celebration. That brochure, including his brief autobiography, has provided much of the material used in this present retrospective to describe departmental events during Shepard's career.

Shepard died at his campus home at 600 Foothill Road in October of 1997, a few months before his 95th birthday. He had been a member of the Stanford family for more than 75 years.

Chapter 5

Robert A. Huggins and the Creation of Materials Science at Stanford

Since his arrival on the faculty in 1954, Robert A. Huggins has served on the materials faculty at Stanford for 65 years. The key roles that he has played at Stanford, particularly in creating today's Materials Science Department and in establishing the Center for Materials Research, make him one of the central figures in the history of materials science and engineering at Stanford. In addition, he initiated research on solid-state electrochemistry at Stanford that eventually led to research on battery materials that became one of the hallmarks of the department in the early part of the 21st century.

In addition to his many Stanford contributions, Huggins also was an early leader of building a national effort in materials research through his work as program manager for the Inter-Disciplinary Laboratory (IDL) program for materials sciences in the Advanced Research Projects Agency of the Department of Defense, and through his work as a co-founder of the Materials Research Society.

Growing Up and Early Days at Stanford

Robert Alan Huggins was born on campus in 1929, the son of the well-respected polymer chemist, Maurice L. Huggins, then an assistant professor in Stanford's chemistry department. Huggins was born in the home that his father built at 600 Foothill Road, a residence later owned by O. Cutler Shepard. Huggins' mother was not able to deliver him at the Palo Alto Hospital on Embarcadero Road because the last bed in the maternity ward had been taken by Sibyl Terman, wife of Fredrick E. Terman, then an assistant professor in electrical engineering (later Stanford's Dean of Engineering and university Provost).

When Huggins was three years old, his father took a sabbatical leave at Johns Hopkins University and soon after accepted an offer to join the faculty there. Huggins started school in Baltimore, but completed his primary schooling in Rochester, New York, where his father had accepted a position at Eastman Kodak. In 1942, young Huggins was enrolled in Deerfield Academy, a boarding school in western Massachusetts known as a top college preparatory school. Graduating in 1946, he went to Amherst College to study physics, then on to MIT, where he studied with John Wulff and Harry Udin in the Metallurgy Department. With fellowship support from International Harvester, he worked on phase transformations in steels at high rates, leading to an M.S. degree for his building of the equipment he subsequently used in his doctoral research. He was granted his Sc. D. degree in 1954 with the

Robert A. Huggins, 1954. (Stanford News Service)

dissertation "The Formation of Austenite in Plain Carbon Steels at High Heating Rates."

Huggins' decision to accept an assistant professorship at Stanford was met with skepticism by his MIT faculty mentors. John Chipman, head of the Metallurgy Department pointed out the obvious: Stanford was not a top department in metallurgy. But Huggins came anyway, perhaps seeing a potential at Stanford that his MIT mentors did not.

Shortly after arriving at Stanford, Huggins was invited to give a seminar to the Chemistry Department, then located in the Old Chemistry Building, in front of the Quad. He arrived a little early to make sure he would be prepared, and a secretary responsible for arranging his visit asked if he would like to have a cup of coffee. Having been with the Chemistry Department when Huggins' father was there in the early 1930s, and having known him as a three-year old boy, the secretary addressed Huggins as "Bobby." Whether that was considered endearing or deflating by the young assistant professor is not known.

Despite the bleak outlook for metallurgy in the 1950s, Huggins and Shepard did their best to promote the field. Huggins's influence on the changing direction of metallurgy at Stanford and its move toward materials science can already be seen in a 1957 promotional brochure, *Metallurgy at Stanford,* which described the exciting prospects for work in that field. The brochure highlights not only traditional achievements of metallurgy, but also the emerging contributions that would be made in electronics and aerospace.

Shepard took a much-deserved sabbatical leave, 1957-58, to work as an engineering specialist in research at Atomics International in Southern California. Teaching duties in physical metallurgy that year fell to Huggins and to Servet A. Duran, who had been appointed as an acting associate professor. Duran was another in a long line of graduate students from Turkey. He had completed his Engineer's thesis in 1946 on the use of x-ray diffraction for the determination of grain size in powdered metals and was to complete his Ph.D. in 1963 with a study of the temperature dependence of creep of metals. That same year, Fred Terman, then serving as the university's provost, asked Huggins to begin a search for another faculty member in metallurgy.

The faculty search initiated by Huggins led to major changes in the department. At this time, the program was then a Division of Metallurgical Engineering in the School of Mineral Sciences. Charles Frederick Park, Jr., a distinguished economic geologist, would serve as its dean from 1950 to 1965. While Park had an outstanding reputation in his field and was much loved as a highly respected teacher, he did not have the kind of hard-driving personality found in his colleague and boss, Provost Fred Terman. Terman was said to be annoyed that Park would not give his responsibilities as dean a priority over his personal teaching and research. Terman instructed Huggins, at the time an untenured assistant professor, to conduct a search for another faculty member in metallurgy.

Terman was well-aware of federal government priorities in science and engineering research, and he likely saw that Stanford would benefit by making a stronger effort in metallurgy. He surely could see that Stanford sorely needed strengthening in that field. As an electrical engineer with many ties to electronics companies growing up in the region, Terman also saw potential in new directions that would attract the interest of local electronics and aerospace companies.

Huggins' search involved the usual advertisements, including letters to major university departments in metallurgy and to relevant government and industrial laboratories. Industrial laboratories were a particularly significant source of academic talent at that time. Huggins reviewed 40 applications, and identified three top people, each of whom, he thought, could be good appointments: David Austin Stevenson, who had been a top chemistry student at Amherst when Huggins was studying physics there and who had received his Ph.D. in Physical Chemistry from MIT. Oleg Dimitri Sherby, who had been a research engineer with John Dorn at U.C. Berkeley and who Huggins knew by reputation; and Victor G. Macres, another Ph.D. student from MIT, but whom Huggins did not know personally. The appointment of Sherby was probably encouraged by Nicholas J. Hoff, head of the aeronautics and astronautics program at Stanford. Hoff had urged Huggins to seek a physical metallurgist who might contribute to the high temperature structures programs in aeronautics.

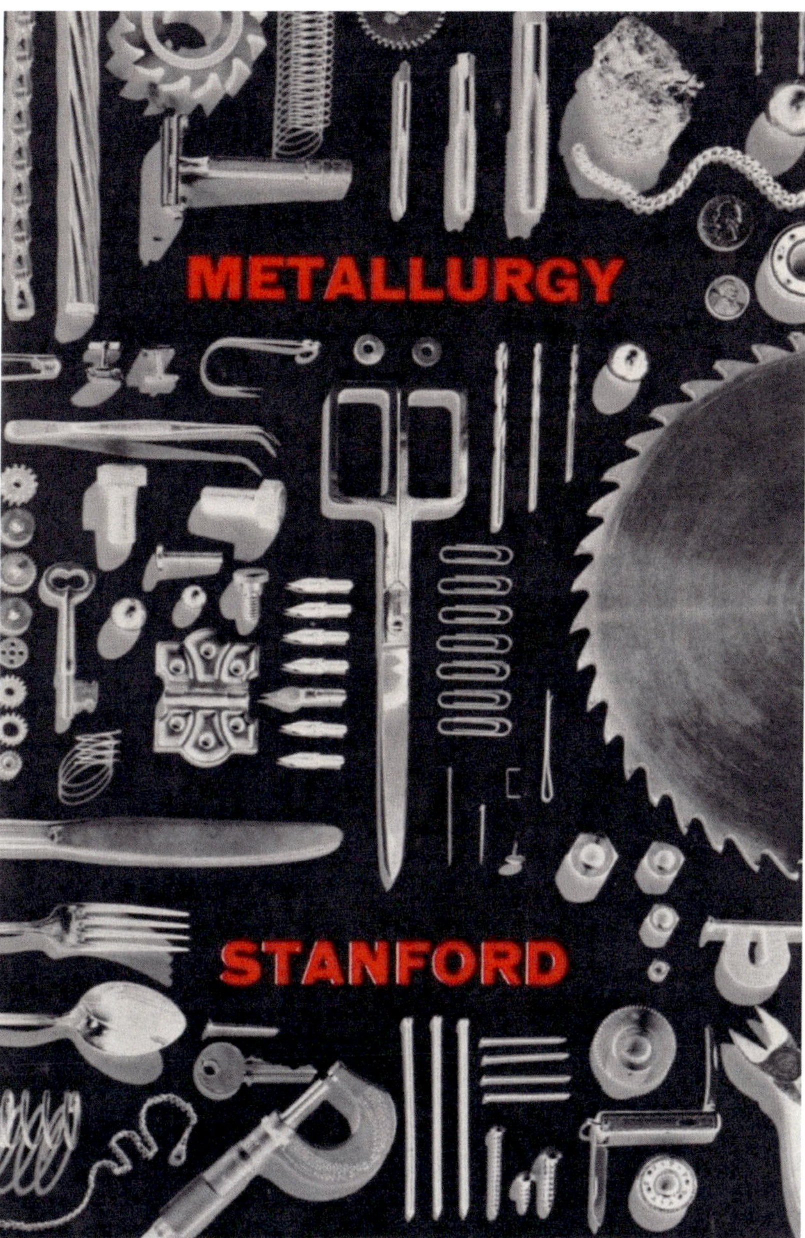

1957 Promotional brochure for Metallurgy at Stanford (MSE Dept.)

As Huggins has said about the search: "I could not decide whom to appoint among these three finalists." He took the files to Terman and explained his dilemma. Terman instructed Huggins to leave the files with him so that he could have a look for himself. A few days later he told Huggins: "Hire them all." Such was the aggressive style that Terman, backed enthusiastically by President J. E. Wallace Sterling, brought to his role as Stanford's provost. Another indication of Stanford's aggressive effort to build up the study of materials around this time are the efforts made to entice Johannes Weertman, David Turnbull, and Ray Smallman to join the materials faculty in each of the Decembers from 1960 to 1962. Although these offers were declined, the high quality of the people being sought for the faculty would be confirmed when each of these individuals was later elected to the highest academies in their respective countries.

Creating the Materials Science Department

Shortly after the arrival of Sherby, Stevenson, and Macres in 1958, Huggins and Shepard were able to create the Materials Science Department. Sometime in 1959, Provost Terman (who had relinquished the position of dean of engineering to Joseph M. Pettit the year before) approached Shepard and Huggins with a proposal to move the department back to the School of Engineering, where it had been from 1925 to the creation of the School of Mineral Sciences in 1947.

Terman could see the growing need for work on materials and wanted that subject linked to other engineering school departments, where it could attract interdisciplinary interest as well as better funding from federal agencies and industry sources. Before a final decision was made, however, President Sterling called Shepard and Huggins to his office. Sterling had worked well with Terman for years and was used to smoothing the edges of Terman's personality when it came to faculty interactions. He asked a simple question: did they want to make the move proposed by Terman? Shepard and Huggins reassured the president that they indeed saw the move to the School of Engineering as a very positive development for the department.

The department moved to the School of Engineering with a completely new name: Materials Science. The Stanford Board of Trustees approved the move at their September 1960 meeting, and the newly named department, with O. Cutler Shepard as executive head, first appeared as part of the School of Engineering in the *Courses and Degrees* bulletin for the 1961 Summer Session.

(As an historical note, Northwestern University was the first university to create a materials science department, having done so formally in 1958.)

Creating the Center for Materials Research

In addition to his key role in establishing the new Materials Science Department, Huggins created the Center for Materials Research (CMR) at about the same time. That center, together with the department, put Stanford on the map in the growing field of materials research.

It can be argued that the spark that catalyzed the creation of the academic field of materials science and the interdisciplinary field of materials research occurred on October 4, 1957; on that day, the Soviet Union successfully launched Sputnik I, the first artificial satellite to successfully orbit the Earth. This event set off alarm bells among those responsible for the national economy and security of the United States.

Soon after the October event, a number of federal agencies convened panels, commissions, and committees to develop recommendations for a coordinated response. Independent studies by the Atomic Energy Commission's Metallurgy and Materials Branch, the Office of Naval Research's Solid-State Sciences Advisory Panel, and the National Academy of Sciences' Committee on Materials (with support from the Air Force) all recommended immediate action to coordinate materials research at the national level. These activities caused President Dwight Eisenhower's President's Science Advisory Committee (PSAC) to also identify materials science as an urgent priority. PSAC's 1958 study, entitled "Strengthening American Science," seconded the conclusions of these federal agencies. While the agencies focused on industrial and government research, PSAC urged that much more attention be given to university research. Too much focus on applied research in industry and government could lead, in the long run, to second rate technology, their report argued, while universities were better prepared to develop creative new technologies. PSAC recommended a significant increase in federal funding to universities to provide long-term support and to help build and equip academic materials research laboratories.

James Killian, President Eisenhower's Special Assistant for Science and Technology, appointed a top-level Interagency Coordinating Committee on Materials Research and Development (CCMRD) to follow up on the PSAC proposals. The agency representatives recognized that broader interdisciplinary education in materials was needed to make the progress expected in materials research. In 1959, they recommended that interdisciplinary laboratories for materials research be built on university campuses across the country to conduct materials research and to train graduate students. CCMRD designated the Advanced Research Projects Agency (ARPA) of the Defense Department (DOD) as the lead federal agency for materials programs. ARPA had been created in 1958 by President Eisenhower to "formulate and execute research and development projects that would expand the frontiers of science and technology beyond the needs of immediate military requirements." (It would later change its name to DARPA, the Defense Advanced Research Projects Agency.)

By Spring 1959, ARPA had created the Interdisciplinary Laboratories (IDL) program for materials research and had begun to visit academic institutions already engaged in significant DOD and Atomic Energy Commission (AEC) work in materials. Stanford was not included in those first visits because it was not, at the time, considered to be a strong enough program in this area. But ARPA, in announcing the initiative, had sought proposals from any interested university. Late that summer, Huggins had learned about the initiative circuitously, from his Office of Naval Research (ONR) contract monitor.

With help from faculty in electrical engineering, physics, chemistry, chemical engineering and materials science, Huggins quickly put together a proposal for funding that included an aggressive equipment list totaling nearly $500,000; this request was later pared down to a more modest $85,000. An equipment grant was made to Stanford in 1960, but the program still was not considered sufficiently strong to compete for one of the first IDL program grants. (Three other universities were selected for IDL programs that first year: Cornell University, Northwestern University, and the University of Pennsylvania.) But Stanford immediately started planning for follow-up materials programs that might be announced later by ARPA, and Terman appointed an interdepartmental committee to oversee that effort, with 31-year-old Huggins as chair.

Huggins learned about subsequent competitions for IDL programs, but almost missed being considered in the second round when the ARPA announcement was misplaced in the president's office. Once the announcement was found, an effort was made to express interest to ARPA in being considered for the IDL program. ARPA representatives visited Stanford in July 1960, and informed Huggins and his committee that if Stanford wanted to be in the IDL program they must submit their proposal before the end of October. Huggins and the others quickly pulled together a proposal, leading to a second ARPA site visit in January 1961.

At the 1961 site visit, Provost Terman gave Stanford's main pitch, using the success of the interdisciplinary Microwave Laboratory as an example. Huggins explained how an interdisciplinary center would meet the objectives of the ARPA program by acting as a catalyst for additional research funding, rather

Robert A. Huggins, in his lab in the McCullough Building. (MSE Dept.)

than a main source of funding for a select few. From the very beginning, CMR was envisioned as a center that would provide seed funding for promising new projects, central facilities for joint use by students and faculty, and a common meeting place for interdisciplinary interactions. The site visit was a success, and Terman, Huggins, and others were invited to ARPA to negotiate a contract the following summer. The result was a $2.6 million contract for four years, making Stanford one of five universities awarded IDL programs in the second round, joining MIT, Harvard, University of Chicago, and Brown University.

Solid-State Electrochemistry and the Seeds of Battery Research

Although Robert Huggins had many graduate students and active research programs in the late 1950s and 1960s, he devoted much of his time getting the Materials Science Department and CMR off the ground. This effort included, among many other things, overseeing the renovation of the Mining and Metallurgy Building, which became the Peterson Laboratory for the Materials Science Department; and overseeing the funding, design and construction of the McCullough building, where CMR was to be housed.

By 1965, with these tasks largely completed, he was able to escape to Germany for a well-earned sabbatical leave. He elected to spend that year with Carl Wagner, the esteemed physical chemist, then Director of the Max Planck Institute of Physical Chemistry at the University of Göttingen. Wagner, considered by some to be the "father of solid-state chemistry," had been a professor of metallurgy at MIT when Huggins was there, and Huggins had taken several courses from him.

On his return from Germany, Huggins immediately began to offer courses in the general area of solid-state electrochemistry. He offered his signature course, MSE 232, "Point Defects in Crystals," for the first time in 1966. The course focused on defect equilibria in non-metallic crystals and the influence of temperature and chemical and electric potentials on the properties of these materials. Huggins would teach that course for 25 years (it was entitled "Solid State Ionics" in the later years). By 1980, Huggins had introduced MSE 204, "Workshop in Energy Storage," which later became a course entitled "Energy Storage."

These courses, and Huggins' burgeoning research in solid-state electrochemistry, would provide the foundation for his well-regarded books, *Advanced Batteries* (2008) and two editions of *Energy Storage* (2010) published by Springer a decade into the new century. His work pre-dated and provided a fundamental underpinning for the explosion of research in the department on batteries, fuel cells, and electrochemical sensors that was to come in the first two decades of the 21st century.

Since Huggins had not been working in the area of solid-state electrochemistry before his sabbatical leave in 1965-66, he had to start from scratch in that field when he returned to Stanford. By the mid-1970s, Huggins had assembled a very strong group of graduate students, post-docs, and research associates. That group included his long-time collaborator Werner Weppner, who had come from the University of Dortmund and would later head the Solid State Electrochemistry Group at Stuttgart before moving to the University of Kiel. After he became an emeritus professor at Stanford, Huggins would go on to spend time with Weppner at Kiel.

Also included in the 1970s group were research associates Bernard Boukamp and Ian Raistrick, and graduate students Chun Ho, John Wen, and Turgut Gür. Gür would later work with Huggins as a senior research associate; he became a consulting professor in the MSE Department through his work with Friedrich Prinz and others on projects involving energy conversion and the environment. M. Stanley Whittingham played a key role in helping the Huggins effort get off the ground in the early 1970s. He would go on to a Distinguished Professorship at the State University of New York at Binghamton and would be elected to the National Academy of Engineering in 2018 for the kind of work he started with Huggins.

The work that Huggins and his group did in the 1970s and early 1980s provided a scientific foundation for the explosion of effort on battery technology that occurred in the early part of the 21st century. Lithium ion batteries are ubiquitous in present day technology, from the smart phones we carry in our pockets to the electric scooters, skateboards, and cars that we now see every day. These batteries are

McCullough Building, 1973. (Stanford University Archives)

Robert A. Huggins, 1992. (MSE Dept.)

outgrowths of the development of new electrode materials capable of storing huge amounts of electric charge which, in turn, allowed development of batteries with unprecedented energy and power densities. Huggins' group can claim to have contributed significantly to these developments.

In a 1977 paper, "Determination of the Kinetic Parameters of Mixed-Conducting Electrodes and Applications to the System Li3Sb," in the *Journal of the Electrochemical Society*, Weppner and Huggins introduced the electrochemical cell method for determining both the thermodynamic and kinetic properties of mixed (electronic and ionic) conducting electrodes that would lay a scientific foundation for research on battery electrode materials for decades to come. The beauty of this technique is that voltage measurements could be directly linked to the electrochemical potentials in the conductors, while current measurements were directly linked to the kinetics of ionic transport. To illustrate the use of that technique, Weppner and Huggins investigated the compound Li_3Sb (lithium antimonide), a mixed-conducting compound with potential for use as an electrode in lithium ion batteries. The titration method they developed allowed them to study the properties of Li_3Sb as a function of composition, even though the range of composition in that compound is extremely small.

This work was followed in the next few years by several papers using similar techniques, also in the *Journal of the Electrochemical Society*. The most frequently cited paper from that group was entitled "Application of AC Techniques to the Study of Lithium Diffusion in Tungsten Trioxide Thin Films" by Ho, Raistrick, and Huggins. In that paper, Huggins and his team used an AC version of the Weppner-Huggins method to determine the ionic diffusivities and thermodynamic properties of lithium-tungsten trioxide as a function of composition. Strong interest in this paper was partly related to the potential applications of lithium-tungsten trioxide in electrochemical display technologies.

The Huggins group published several other papers on the kinetic and thermodynamic properties of lithium phases, including the LiAl compound and an all-solid lithium electrode consisting of finely dispersed LiSi phases in a mixed conducting matrix of LiSn.

In 1981, Wen and Huggins published another important paper, entitled "Chemical Diffusion in Intermediate Phases in the Lithium-Silicon System" (*Journal of Solid State Chemistry*). Wen and Huggins used the equilibrium titration method to study chemical diffusion in each of the intermediate phases in the Li—Si (lithium-silicon) alloy system. As with earlier studies, they were able to examine the chemical diffusivity of lithium as well as the thermodynamic properties of the intermediate phases, in spite of the narrow range of compositions exhibited by these compounds. The 1981 Wen-Huggins paper would become extremely important decades later when serious attention was given to using silicon as a positive electrode in lithium ion batteries. It had been known since the mid-1970s that silicon had the capacity to store huge amounts lithium (about four lithium atoms for every atom of silicon) and thus had great potential as an anode. But storing that much lithium in silicon is accompanied by a volume expansion of the silicon electrode of 300 percent. Subjecting almost any crystalline material to that kind of volume expansion causes it to fracture and crumble, a failure process now called decrepitation. This failure process prevented the use of silicon as an electrode for lithium ion batteries.

In 2008, Candice Chan, working with Yi Cui and Huggins, found that by using as the anode silicon nanowires below a critical diameter, the volume expansion could be accommodated by the spaces between the nanowires. The Chan et al. paper, "High-Performance Lithium Battery Anodes Using Silicon Nanowires," published in *Nature Nanotechnology*, has become the most frequently cited paper of any of the seven co-authors, with over 5200 citations by 2019, according to *Google Scholar*.

National Leadership of Materials Research

Even as Huggins was focusing his personal research on solid-state electrochemistry and potential battery materials throughout the 1960s and 1970s, he continued to serve in leadership roles in research at Stanford and nationally.

Huggins had been the Director of the Center for Materials Research (CMR) at Stanford since it was founded in 1961, under the Interdisciplinary Laboratory (IDL) program of the Advanced Research Projects Agency (ARPA) of the Defense Department. When the post of Program Manager for the IDL program became vacant in 1969, Huggins was chosen to lead the federal program because of his long experience in materials research and his knowledge of the operations of the federal funding of research. He assumed that position in 1969 and was faced with an immediate existential threat to the continuation of the program.

All through the 1960s there had been questions about whether the IDL program and other basic research programs like it were conducting research that was relevant to the needs to the Defense Department. The issue came to a head in 1969 when Senator Mike Mansfield of Montana introduced an amendment to the Military Procurement Authorization Act which effectively banned the use of the DOD budget to fund "any research or study unless such project or study has a direct and apparent relationship to a specific military function or operation." Since the IDL program was fully funded by the DOD, it was squarely in the crosshairs of the Mansfield Amendment.

Huggins immediately organized an effort to show that the research being conducted in the IDL program was indeed relevant to the long-term needs of the Defense Department, but the mood in Congress was not particularly sympathetic to appeals from universities where students were protesting the activities of the Defense Department in prosecuting the Vietnam War. In spite of Huggins efforts, it became apparent in late 1970 that the DOD could no longer justify continued funding. It fell to Huggins to find a solution to this critical national problem.

Through his various contacts in Washington D.C., he managed to get the IDL program transferred to the National Science Foundation in 1972 under a new name: the Materials Research Laboratory (MRL) program. The MRL program provided funding for materials research until the mid-1990s, when it was replaced by other NSF funding mechanisms. Thus, Huggins' efforts in late 1970 and 1971 resulted in more than 20 years of continued national funding for materials research.

Having been the manager of the DOD-IDL program in the Materials Sciences and having

Robert A. Huggins (R. Huggins)

successfully negotiated its transfer to NSF as the Materials Research Laboratory program, Huggins became a national leader in the burgeoning multi-disciplinary field of materials research. It therefore was natural for him to be included in the group of people who met in 1973 to establish the Materials Research Society (MRS). Much as the Center for Materials Research was a multidisciplinary activity involving faculty and students from many different disciplines, so too would MRS have a similar character. As described in a separate section below, MSE faculty and former students have played a major role in making MRS the dominant professional society for the field of materials science and engineering.

Now in his 91st year, Robert Huggins continues to come to the department on most days, maintaining his interest in all aspects of materials science, and most especially energy storage.

Part III ~ Building a National Reputation

Chapter 6

Nucleation of a Nationally Prominent Department

The academic year 1958-59 was a seminal moment in the history of the department. While still organized as a Department of Metallurgical Engineering in the School of Mineral Sciences, the seeds were being planted for the Materials Science Department that was to come. Three new faculty members--Oleg Sherby, David Stevenson, and Victor Macres—joined Cutler Shepard and Robert Huggins that year, beginning the department's transformation into a nationally prominent materials program. Each of the three brought new ideas and ambitions that set the tone for the explosive growth of the 1960s.

Of the three, Sherby was the most senior and experienced, having worked as a research engineer with John Dorn at Berkeley for nearly a decade. He was appointed as an associate professor, the same rank that Huggins then held. Stevenson and Macres were appointed as assistant professors. With this faculty expansion and the dramatic increase in faculty that soon followed, facilities for work on materials also improved substantially. A major renovation of the old Mining and Metallurgy Building in 1962, was followed by construction of the McCullough building, a new facility for materials research, a few years later.

Oleg D. Sherby ~ High Temperature Deformation and Superplasticity

Oleg Sherby already was well known for his work on high temperature creep of metals when he arrived in 1958. His reputation in that field would grow exponentially during his time at Stanford, however, and he was to become one of the giants of materials science. His work led to his being the first regular faculty member in the department to be elected, some 20 years after his arrival, to the National Academy of Engineering.

Oleg Dimitri Sherbynin (Sherby) was born in Shanghai, China in 1925 to Russian parents who had fled Russia a few years earlier to escape the Russian Revolution. Sherby used to say that in 1924 his father had walked 200 miles to get to a railroad station in China so that he could be reunited with his wife, then in Shanghai. Sherby spent his early childhood in Shanghai and had vivid memories of it. On a return trip in the 1980s, he found the apartment building where he had lived in the early 1930s, and had someone take his picture with a bicycle in the very same location and with the same pose as in an old photograph of him taken there when he was a seven-year-old boy. He enjoyed showing the two photographs side-by-side, a very Sherby-esque thing to do.

Oleg D. Sherby, 1958. (Stanford News Service)

In the late 1930s, Sherby's family had to move again, this time to flee the Japanese bombing of Shanghai. That move brought him to the San Francisco Bay area, where he finished high school. He enrolled at U.C. Berkeley, where he planned to study chemical metallurgy, but his studies were interrupted by World War II and service in the U.S. Army Infantry and Corps of Engineers. After his honorable discharge in 1946, he returned to Berkeley, but changed his major to physical metallurgy, eventually earning both undergraduate and Ph.D. degrees there.

While completing his Ph.D., and for several years thereafter, Sherby worked as a research engineer at Berkeley for Professor John Dorn, who had attracted a number of grants and contracts for the study of creep of metals. A summary of his work in this area can be found in the memorial biography prepared by Jeffrey Wadsworth and William Nix following Sherby's death in late 2015.

His early reputation was built on his discovery of the intimate relation between lattice self-diffusion (the movements of individual atoms) and high temperature deformation of crystalline materials. Before his time, the important problem of creep, the slow deformation of metals at high temperatures, had naturally received much attention because of its relevance for gas turbine engines, nuclear reactors and other generating systems. But the atomic processes controlling creep had not been identified. In the early 1950s, when he was working as a research engineer at U.C. Berkeley with John Dorn, Oleg began to draw correlations between the rate of high temperature creep of different metals and the rate of self-diffusion in those same metals. He soon discovered that for a wide variety of metals the rate of creep could be accurately predicted with knowledge of the self-diffusion alone and a few other physical properties. It was only after this that theorists started to catch up and find the microscopic reasons for the correlations that Oleg had found. By the mid-1960s he had developed a very complete phenomenology for high temperature creep of metals that served not only as a guide for designing heat resisting alloys but as a solid body of facts about high temperature creep to which modern theories must conform.

Oleg was a master at developing phenomenological relations between different physical properties of materials. He may have been inspired by Trouton's rule, a phenomenological rule stating that the entropy of vaporization for all liquids is nearly the same at their boiling points. His finding that the rate of steady state creep of metals is directly proportional to the rate of lattice self-diffusion and that the seemingly unrelated properties like high temperature strength of metals could be predicted accurately from a knowledge of atomic self-diffusion is an example of his mastery of phenomenology. He was fond of saying that he had found "all of the data in the world" in reaching his conclusions.

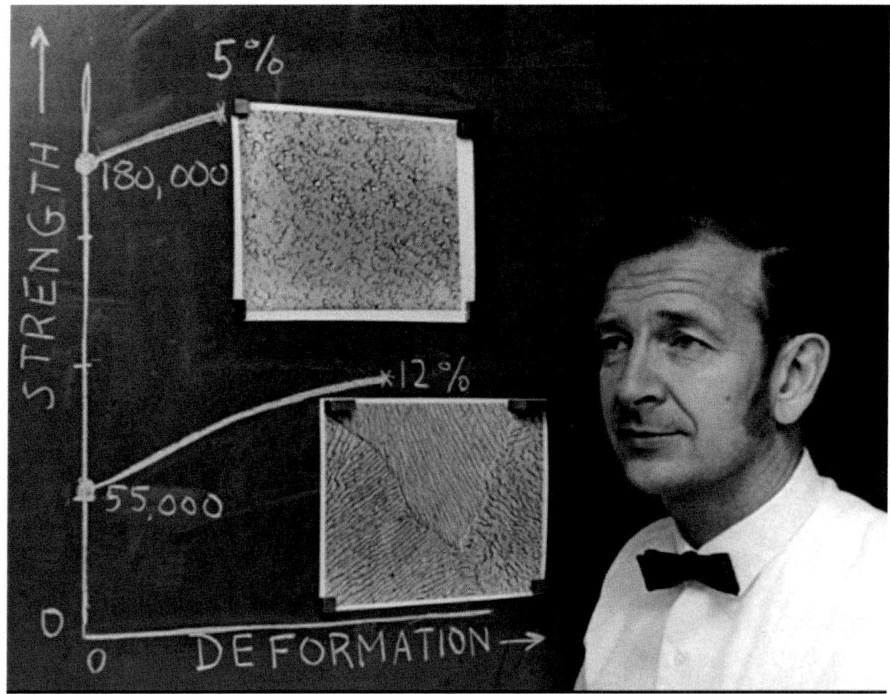

Oleg Sherby, teaching in 1969. (Stanford News Service)

> Indeed, by finding "...all the data..." he was able to develop his impressive account of high temperature creep of metals that has stood the test of time and has led to the development of new alloys. *(NAE Memorial, 2017)*

Sherby quickly built a research group at Stanford and continued to focus on identifying the factors that control the strength of solids at high temperatures. One of his seminal papers in 1962, "Factors Affecting High Temperature Strength of Polycrystalline Solids" (*Acta Metallurgica*), became a cornerstone of his reputation. In it, he showed that the rate of high temperature deformation of crystalline solids depended not only on the lattice self-diffusion coefficient, but also on the crystal structure and the temperature dependent elastic modulus.

Sherby had a number of students measuring the elastic modulus of metals as a function of temperature and showing the phenomenological connection to high temperature strength. Later, through his work with Craig R. Barrett, one of his students, he showed that the stacking fault energy for face centered cubic metals was another factor to be included in the developing phenomenology. By the late 1960s, this phase of his work on the fundamentals of high temperature creep was essentially finished and he summarized all of his findings in a 1968 review article in *Progress in Materials Science*, which, along with the 1962 paper, are his two most highly cited works.

Although Sherby's focus on high temperature deformation started at Berkeley with John Dorn, he was also following up on previous studies at Stanford by Shepard and Huggins, both of whom had students working in that broad area of metallurgy and materials science in the 1950s. Sherby inspired younger faculty members William Nix and Craig Barrett to also take up work on high temperature deformation, leading to what some called the "Stanford School of High Temperature Deformation." This was but one branch of the group nicknamed the "California School of High Temperature Deformation," started by John Dorn; the other branches were located at U.C. Davis and USC.

Later in his career, Sherby turned his attention to the phenomenon of superplasticity. Again, quoting from the NAE memorial biography:

> In the late 1960s he was one of the first to explore the phenomenon of superplasticity, the ability of some metallic alloys to be stretched several times

This photo of Oleg Sherby, with a 17th century Persian saber, illustrated the June 1992 **Los Angeles Times** *story, "Super Steel: The Stuff of Legend?" (L.A. Times)*

their initial their lengths without breaking or, as Oleg would say, "Like well-chewed chewing gum." He showed how these properties could be used in metal forming and was soon leading that field by identifying the various ways that alloys could be made superplastic. Subsequently, a race began to demonstrate this property in steel. He determined that superplasticity could be developed in steel by raising the carbon content to very high levels where conventional wisdom held that such compositions were impractical. The steel families he developed, now called ultra-high carbon steels, could not only be made superplastic, and thus formable by the right kind of processing, but they also had remarkable room temperature strength and ductility. His work on superplasticity subsequently extended to certain ceramic materials, which are often assumed to be completely brittle. His work on ultra-high carbon steels also led to the discovery that they had similar compositions to the famous steel swords of Damascus. This led to a rediscovery of how the ancient patterns on the swords were made. This stimulated related research into other ancient laminated steels and their similarity to contemporary materials. This aspect of Oleg's work was described in a 1981 article in the *New York Times*. The article described how the Stanford team's modern methods produced the same carbon-rich steel used during the Crusades.
As *Times* science writer Walter Sullivan wrote, "Swords of this metal could split a feather in midair yet retain their edge through many a battle." *(NAE Memorial, 2017)*

Sherby's enthusiasm for life was reflected in his passion for sports and athletics. He ran track and played soccer at Berkeley, and at Stanford he was remembered for organizing noon volleyball games for faculty and students alike. It was not uncommon for him to get everyone out to the volleyball court at least twice a week. Even distinguished international visitors to the department were encouraged to join in, so that the rosters included some of the Who's Who of materials science. He was, likewise, an enthusiastic and frequent participant in the after-work ping-pong games that students organized in the department.

Sherby also was a very good teacher. He thought that his teaching scores were better when he wore a tie, so he regularly wore one when he

Oleg Sherby and Conrad Young, 1969. (Stanford University Archives)

taught. But it was not the tie—it was his warm, enthusiastic personality that made him exceptional. Anyone who met Sherby will remember his *joie de vivre*. He maintained his enthusiasm and interest in research throughout his life. He was thinking about ultra-high carbon steels and was working on a new approach for understanding these materials at the time of his death in 2015.

An award in Sherby's memory was established by The Minerals, Metals & Materials Society (TMS). The Oleg D. Sherby award recognizes an individual, (or small group of collaborators), who has (have) made significant contributions to the understanding of the behavior of materials at high temperatures. The award includes a cash prize as well as a medal and a certificate. In addition, TMS created the Wadsworth-Sherby Bladesmithing Grand Prize, in honor of the work Sherby did on Damascus Steels with his longtime colleague and associate, Jeffrey Wadsworth. The award is given for the best work in a student-led competition to make a knife. The award was given for the first time in 2017.

David A. Stevenson ~ So Close to a Nobel Prize

With his undergraduate and graduate degrees in chemistry and post-doctoral work in metallurgy, David Stevenson was ideally prepared to bring the teaching of thermodynamics and electrochemistry to a new level in the department. He was responsible almost singlehandedly for training Stanford materials science students in these subjects for 35 years. In addition, soon after he arrived in 1958, he introduced the first research on electronic materials, a focus that came to dominate the attention of the department for several decades. His work on electronic materials with one of his students, Herbert Maruska, brought him closer to sharing a Nobel Prize than any other person in the department's history, thus far.

David Austin Stevenson was born on September 6, 1928, in Albany, N.Y. In 1946,

David A. Stevenson, 1958. (Stanford News Service

he graduated from Albany Academy, a school that had been attended by Joseph Henry (of electromagnetic induction fame) and Herman Melville (of Moby Dick fame). He showed promise for a career in science by winning the senior prize in physics. He went on to study chemistry at Amherst College, where he was friends with Robert Huggins, who was studying physics there. Stevenson graduated *summa cum laude*, in 1950, with a B.A. in Liberal Arts, moving on to MIT for graduate work in chemistry and his 1954 physical chemistry Ph.D. degree. Following a year as a Fulbright Fellow at the University of Munich, Stevenson accepted an appointment as a research associate in the Metallurgy Department at MIT, working with Professor John Wulff, one of Huggins' doctoral advisors. After three years working with Wulff on the solubility of metals in molten lead, Stevenson moved to Stanford as an assistant professor. Stevenson's first paper at Stanford was on phase equilibria in the silver-antimony-tellurium alloy system, published with his first student, Robert A. Burmeister, Jr. That led to other papers, with Burmeister and others, on the compound semiconductor cadmium-selenide (CdSe), a material that was later to find applications in optoelectronic devices, laser diodes, biomedical imaging, and nanosensing. His work in this area introduced electronic materials into the department and resulted in several of his students becoming leaders of this field. Burmeister went on to leadership positions at Hewlett-Packard Labs, while another student, Richard A. Reynolds, became a recognized authority in the field of microelectronics. Reynolds' career included work at Texas Instruments and a position as vice president and technical director of the Hughes Research Laboratories, but his greatest impact came through his long tenure as director of DARPA's Defense Science Office, where he sounded the alarm about the sparse federal funding for growth of electronic materials. Reynolds was credited with promoting the explosion of research using molecular beam epitaxy (MBE) for growing compound semiconductors for optoelectronic applications. He was the recipient of the first Welton J. Crook award in 1960.

David A. Stevenson, 1979. (Stanford News Service)

In the fall of 1970, one of Stevenson's new graduate students, Herbert Paul Maruska, arrived at the department from the RCA laboratories in Princeton, NJ. He came with extensive experience working on the growth of compound semiconductors. Working with James J. Tietjen, a young chemist at RCA, he had been asked to develop a process for growing thin films of gallium nitride (GaN). Tietjen had postulated that blue light-emitting diodes (LEDs) could be created using gallium nitride. He and others knew that such a development would lead to white light by mixing the blue with existing red and green LEDs, and that white light LEDs would make flat panel televisions possible, ending the use of cathode ray tubes for that purpose. After several years of development, Maruska was able to grow gallium nitride with various dopants. Based on that work, he was granted a David Sarnoff Award to come to Stanford for doctoral work specifically to develop a bright blue GaN LED. Jacques Pankove and Edward Miller at RCA continued to develop blue LEDs, in their case based on Zinc (Zn) doped GaN.

The daunting challenge Maruska had been given by his superiors at RCA can best be appreciated by jumping ahead more than 40 years. In 2014, the Nobel Prize in physics was awarded for the development of the blue LED. In announcing the prize, the Nobel committee stated:

> When Isamu Akasaki, Hiroshi Amano and Shuji Nakamura produced bright blue light beams from their semiconductors in the early 1990s, they triggered a fundamental transformation of lighting technology. Red and green diodes had been around for a long time but without blue light, white lamps could not be created. Despite considerable efforts, both in the scientific community and in industry, the blue LED had remained a challenge for three decades. (*Nobel press release, 2014)*

As reported in a 2015 *Solid-State Electronics* paper by Maruska and his colleague, Walden C. Rhines, the development of blue light-emitting diodes (LEDs) had a very long and rich history extending back nearly a century and involving many different people, organizations, and technical developments. The group at RCA, and Maruska himself, had come very close to achieving the objective of creating a bright blue LED in the early 1970s. In addition to providing a comprehensive review of this field, the 2015 Maruska-Rhines paper gives a good account of the 1970s effort that Maruska made to develop a blue LED as a part of his dissertation work. Here is a relevant paragraph from the 2015 paper:

> Aware that a Ph.D. thesis has to be based on original research, Maruska realized that he needed to adopt a different approach compared to his colleagues back East. He spoke at length with another graduate student, Walden C. Rhines, who sat at an adjacent desk in the McCullough Building at Stanford. They decided to substitute magnesium for zinc to create a novel acceptor. Soon they were preparing GaN films doped with magnesium which were pale yellow in color and electrically insulating. In early June of 1972, several device structures were prepared based on a thick undoped (n-type) GaN film topped with a thin Mg-doped insulating film. Point contact probes were used for providing electrical contact. On June 8, 1972, after applying 150 volts to a set of point contacts, the first example of violet light emission from Mg-doped GaN was observed. The emission peaked at 425 nm, which is indeed in the violet region of the visible spectrum. The human eye has very low sensitivity in this wavelength range, so such devices did not appear to be very bright, and in fact, looked blue. But they could be easily seen even with the room lights on. *(Maruska-Rhines, 2015)*

Due to contamination, the blue light-emitting diodes first reported by RCA and Maruska in the 1970s were never efficient enough for commercial products. Devices created by metalorganic vapor-phase epitaxy, which was developed much later, as well as other developments, would be needed to produce the very bright blue LED for which the Nobel Prize ultimately was given. Still, with their work on the blue LED, Maruska, Stevenson, and Rhines came as close to sharing a Nobel Prize as anyone from the department ever has.

In addition to having a productive academic career, Stevenson was another department faculty member with a lifelong love affair with athletics and athletic competition. He had played soccer and tennis in high school and was talked into joining the Amherst swimming team by the coach. He lettered in swimming all four years, captained the swim team as a senior, and once held the national collegiate record in the 440 freestyle. At Stanford, he was active in backpacking, hiking, rock climbing, surfing, skiing (he was a National Ski Patrol volunteer in the Sierras), running, biking, and swimming. He was frequently seen scaling the outside surfaces of the Peterson Building like a real-life Spider-Man.

In the early 1970s, Stevenson ran in six marathons and in the 1980s began to focus on the triathlon. From 1986 through 1990, he participated in the five Ironman Triathlon World Championships in Hawaii: a 2.4-mile swim, followed by a 112-mile bike race, followed by a 26.2-mile marathon run. Stevenson won

David Stevenson, Iron-Man Competition, 1990. (MSE Dept)

Bruce Liebert (left) holding a GaN blue (violet) light emitting diode that had been created by Herb Maruska (right), arguably the first blue LED made. (B. Liebert)

three of the competitions in his age group and was always among the top five finishers in his group. He recorded his best time at age 61! It is no exaggeration to say that Stevenson was the consummate scholar-athlete. It was cruel irony that he was to die of mesothelioma at age 65 in 1994.

Victor G. Macres – Electron Optics and Materials Characterization

Victor Gregory Macres was the one who brought materials characterization to the department through the use of electron optics. He had studied with Robert E. Olgivie at MIT and, in his dissertation work, had modified an RCA electron microscope to create one of the first electron probe micro-analysis instruments. Electron probe x-ray micro-analysis involves shining a highly focused electron beam onto the surface of a material and determining the chemical composition of the material by analyzing the x-rays that are emitted. Macres used that technique and his new instrument for his Ph.D. dissertation entitled "Application of Electron Probe Micro-Analysis to Copper-Zinc Diffusion," granted by MIT in 1958. (While Macres and Olgivie were early pioneers in the development of electron probe micro-analysis, they were not the first to build these instruments. In 1951, R. Castaing, studying with the famous x-ray physicist, A. Gunier, at the University of Paris, had invented the first electron probe x-ray micro-analysis instrument as a part of his dissertation.)

When Macres arrived at Stanford in 1958, he was initially assigned the task of teaching courses in physical metallurgy and steel making, but by 1960 he was able to teach courses more suited to his research interests: "X-ray Metallography and Advanced X-ray," and "Electron Metallography." While passionate about electron optics, x-ray diffraction, and electron probe micro-analysis, Macres was a somewhat distracted teacher. It soon became evident that he was more interested in developing a commercial electron probe micro-analysis instrument than in teaching or research. By the early 1960s, he had formed a company called Materials Analysis Company (MAC) in Palo Alto to exploit the advances that he and others had made in electron optics. He continued to teach and attract some students in the early 1960s, but as the development of his company took more and more of his time and attention, it became plain that he could not simultaneously grow his company and keep up his teaching duties. Macres left Stanford in 1965 to devote full time to his company.

While at Stanford, Macres attracted a number of students who were also passionate about electron probe micro-analysis. Richard C. Wolf, who had received his M.S. degree in 1963, worked with Macres after his graduation and later co-authored a book chapter with Macres entitled "Quantitative Microprobe Analysis: A Basis for Universal Atomic Number Correction Tables," for the 1969 book *Electron Probe Microanalysis*. The editor, L. (Bill) Marton, is the same person who collaborated with O. Cutler Shepard in the 1940s in the teaching of "X-rays and Related Methods in Metallurgy." At about the same time, Macres and Wolf presented a paper at the 26th Meeting of the Electron Microscopy Society of America entitled "A combined SEM/Electron Microprobe Analyzer," in which they described the features of the Materials Analysis Company Model 400S. They indicated that the instrument:

> was designed to incorporate the most advanced features of a high-performance electron microprobe analyzer with those of a medium resolution (100 nm) SEM. Now that the SEM was established as a valuable tool, manufacturers such as the Materials Analysis Company were looking for ways to enhance its analytical power by adding existing detectors, like the electron microprobe.

Through the success of his company, Macres was starting to be recognized professionally for his development of the electron probe instrument. He, together with Marton and others, were involved in forming the Electron Probe Analysis Society of America (EPASA), which was the name of the new society until 1974 when it was changed to the Microbeam Analysis Society (MAS).

Unfortunately, Macres did not live long after he had started to make his name in the field. He died at the age of 41, on December 31, 1972. In his memory, the Electron Probe Analysis Society created The Victor G. Macres Award in 1973 to "honor the memory of Dr. Victor Macres, one of the pioneers of electron probe analysis, a dedicated teacher, and a competitive instrument manufacturer." This award is given for the best instrumentation paper presented at the annual meeting of the society and has been supported by Hitachi Scientific Instruments.

Another measure of the esteem in which Macres was held by his professional colleagues is found in a remark made by R. Castaing, in his chapter "Early Times of Electron Mircoprobe Analysis," in the book *Electron Probe Quantitation* (1991). In comparing Macres with another early pioneer, he noted that "the charming little microprobe [Macres] built during the sixties was a model of ingenuity."

New Facilities for Materials Science

The early 1960s was an exciting time for materials science at Stanford. From 1959 onward, the faculty rapidly expanded and new opportunities came with the ARPA funding of CMR. In 1960, the School of Engineering secured

One of the first electron microprobes in the world was built by Victor Macres in the Peterson Building. This instrument was a prototype of the commercial instruments created by Materials Analysis Company in Palo Alto. (Stanford University Archives)

Provost Frederick E. Terman, Mrs. Thomas F. Peterson and Dean Joseph M. Pettit at the dedication of the Thomas F. Peterson Laboratory. (School of Engineering Newsletter)

a major grant from the Science and Engineering division of the Ford Foundation for faculty development in the school. This was followed in 1961 by a supplementary grant for the specific purpose of strengthening faculty and increasing educational opportunities for advanced students working in chemical engineering and materials science.

That funding, together with support from the National Science Foundation and the Thomas F. Peterson Foundation, allowed for a renovation of the L shaped part of the old Mining and Metallurgy building. The original building (which at some point had been numbered 550) was then named the Thomas F. Peterson Laboratory. Peterson had been a graduate of the Electrical Engineering Department in the 1920s and an executive at the American Steel and Wire Company, a division of U.S. Steel, before founding Performed Line Products Co. in Cleveland, Ohio. At the 1963 dedication of the renovated building, Dean Joseph M. Pettit described what the renovation had entailed. He said:

> I should like to call to your attention several significant aspects of this new building. From the outside it would seem not to be a new building at all, but rather a part of the original Stanford Quad with its distinguishing unity of architecture, its sandstone walls and tiled roofs for which Stanford is justly famous. We have seen fit to preserve this exterior, and on the inside to construct an entirely new building. The old interior was what you would expect of architectural styling at the turn of the century. Frame construction, high ceilings, inadequate illumination, and a general wastage of interior volume were to be found here, worsened by the wear and tear of sixty years of occupancy. As you have now seen or will be able to see from a visit through the building, we have replaced the old interior with an entirely new structure, providing modern laboratories and offices together with a great increase in the amount of usable area. Thus, we now have, in effect, a new permanent building with great flexibility for changing utilization. *(Pettit, 1963)*

While the renovated building was a great improvement over what had existed before, it was in no way gold-plated. Limited funding, together with priorities set for the renovation, meant that while there was funding for office furniture for the faculty and staff, laboratories and office spaces for graduate students were essentially empty shells with no new laboratory or office furniture. The still-useful war surplus desks and tables that had been used for years continued their service. The frugality of this approach surely came from the top, from Fred Terman and Robert Huggins, who were never inclined to waste money on frills when there was important work to do.

The need for more space for materials research became even more urgent after the 1961 ARPA contract for CMR was won. In addition to the growth of the materials science faculty, an increasing number of faculty from other academic departments were brought into materials research through CMR. The ARPA contract provided about $1 million for bricks and mortar, but that was insufficient to create the kind of state-of-the-art facility the faculty had in mind. Figuring that the new activities in materials research would have a positive impact on local industry, Huggins and his colleagues hoped to interest a local entrepreneur to fill the funding gap.

Provost Terman had been eyeing Jack A. McCullough, cofounder, with William Eitel, of Eitel-McCullough, one of the oldest electronics firms in the valley, as a potential donor. Eitel-McCullough already had a connection to the Materials Science Department: Leonard Reed, the manager of their Device Research laboratory, had been loaned to the department in 1962 to teach courses on ceramics. Despite some periodic fretting about how the Stanford faculty spent too much time consulting with his competitors, Jack McCullough eventually succumbed to Terman's persuasion and donated $1.5 million to be used for the construction of what would become the Jack A. McCullough Building for Materials Research.

The new McCullough building, completed in 1965, was fully outfitted with specialized facilities for the synthesis, processing, and characterization of modern materials. The generous funding allowed all faculty and student laboratories and offices to be well appointed with new furniture. Some, but not all, of the Materials Science faculty and students were given laboratory and office space in the new facility. By design, the rest of the space in the McCullough building was assigned to faculty research groups engaged in materials work in other departments, such as Electrical Engineering, Applied Physics, Physics, Chemistry and Chemical Engineering. That set the tone for interdisciplinary interactions and collaborations that ARPA had meant to encourage all along.

The Jack A. McCullough Building for Materials Research, 1974.
(Stanford University Archives)

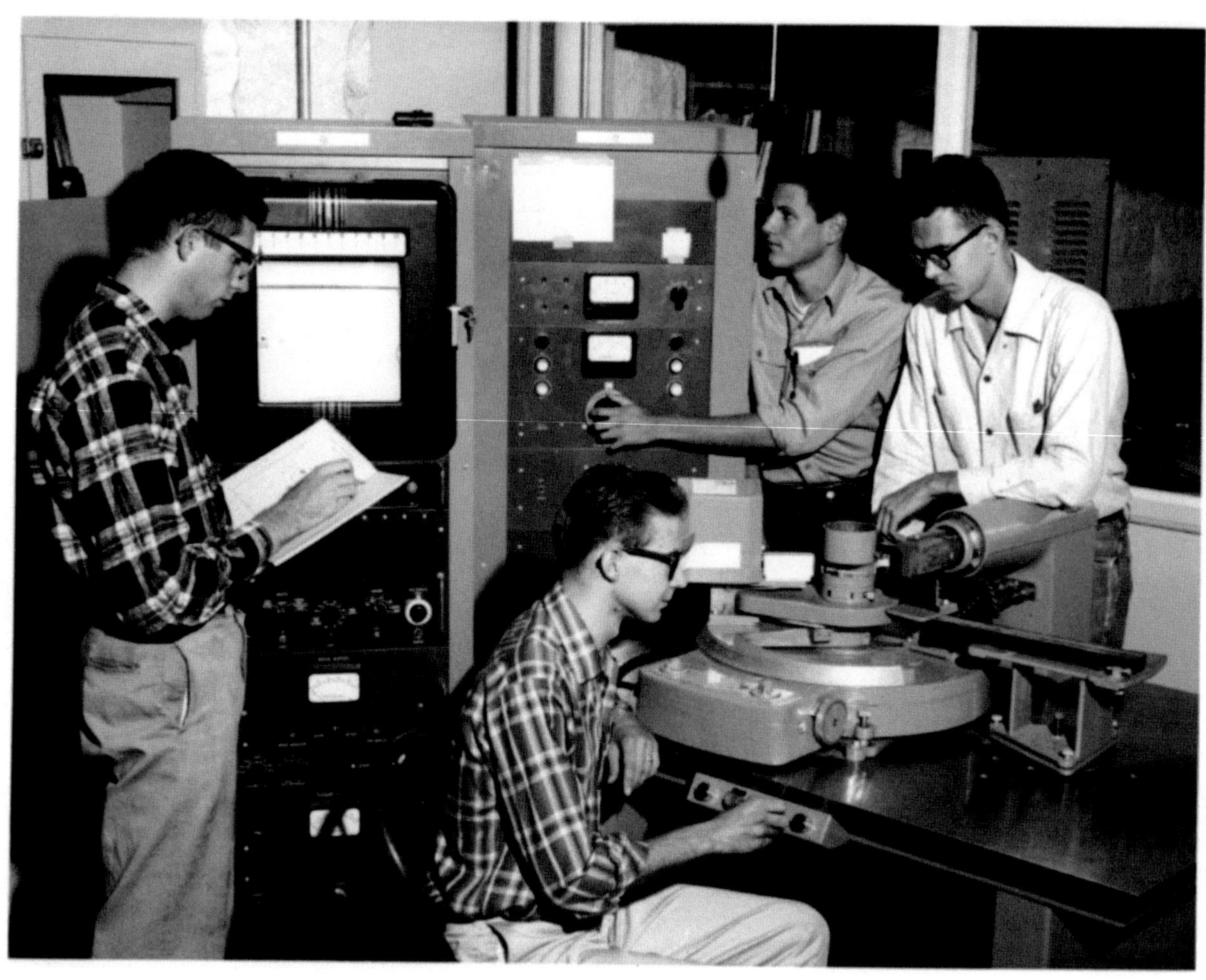

Undergraduate students doing x-ray diffraction experiments in 1959 as a part of an NSF-sponsored summer research program. From left to right: Jim Anderson, Ken Ralls, Richard Gregory and Jerry Coquin. The program was supervised by O. Cutler Shepard and Victor G. Macres. (Stanford University Archives)

Chapter 7

Appointments of Future Department Chairs

After its faculty had grown to five in 1958, moved to the School of Engineering in 1960, and acquired new facilities for materials research, the Department of Materials Science was set for a period of extraordinary growth.

Eight additional faculty members with primary appointments in materials science were appointed between 1960 to 1968: John Cornelius Shyne (1960), Richard Howard Bube (1962), William Dale Nix (1963), William Arthur Tiller (1964), Alan Stephen Tetelman (1964), Craig Radford Barrett (1965), Guy Marshall Pound (1967), and Arthur Irwin Bienenstock (1968). Four of these people (Shyne, Tiller, Bube and Nix) would later chair the department. Here we review the contributions of three of these future department chairs in the order in which they served as chair: Tiller, Bube and Nix. Shyne's contributions are reported in Chapter 8 with those of Tetelman and Pound, who jointly founded Failure Analysis Associates (later, Exponent, Inc.). Barrett and Bienenstock are featured in Chapter 9.

In addition to the regular faculty appointments in the 1960s, three others, with primary appointments in the departments of Electrical Engineering (William E. Spicer and Robert L. White) and Applied Physics (Theodore H. Geballe), received joint appointments in Materials Science. As a result, by 1968, after Macres had left the faculty and Shepard had retired, the regular faculty head count had reached eleven. With two other appointments and one departure over the next three years, the size of the faculty reached twelve and held steady at that point for another decade.

William A. Tiller ~ A Senior Appointment in Solidification Science

In a continuing effort to build up materials science at Stanford, the department appointed William A. Tiller as a professor in 1964. He was the first person to be appointed in the department at the full professor rank and was already well known and respected for his work in solidification science over the previous decade. After receiving his Ph.D. degree from the University of Toronto in 1955, he had spent nine years at the Westinghouse Research Laboratory in Pittsburgh, during the heyday of fundamental research at industrial laboratories in the United States. (Other industrial laboratories leading in fundamental materials research were the GE Research Laboratory in Schenectady, New York; the U.S. Steel Fundamental Research Laboratory in Monroeville, Pennsylvania; the AT&T Bell Laboratories in Murray Hill, New Jersey; the IBM Research Laboratory in Yorktown Heights,

William A. Tiller, 1960s.
(Stanford News Service)

New York; and the Dupont Central Research Laboratory in Wilmington, Delaware.)

Tiller had built up an active research group at Westinghouse. His work there had been supported in part by the Air Force Office of Scientific Research (AFOSR), and he was able to bring a large AFOSR contract with him to Stanford, as well as several staff. Sydney O'Hara came as a senior research associate and was in charge of the experimental parts of Tiller's research. Two engineers who had worked with him at Westinghouse, Gerald Kotler and Lemuel Tarshis, came as doctoral students in the department. In addition, Tiller brought two technicians to support his experimental program.

The necessity of a large federal contract to support Tiller's work was related both to the number of people he brought west with him and to the faculty salary offset policy of the School of Engineering at the time. According to that policy, faculty members were expected to "offset" 50 percent of their academic salary and all of their summer salary with research grants/contracts. A total of 62.5 percent of their annual salary thus had to be covered by research funding. The 50 percent offset policy had been established in the 1950s by then dean of engineering Frederick E. Terman, as a strategy for building up the engineering faculty within the constraints of both limited university general budget funding and a fixed number of university-funded billets. With Stanford's aggressive and highly successful fundraising efforts, begun in the early 1960s, the policy has been relaxed over the years to a point where, by 2019, nearly all academic-year salaries in engineering are provided for by the university budget.

In 1964, Tiller's AFOSR contract for about $600,000/year was by far the largest contract in the department. In 2019, a contact of that size would amount to annual funding of around $5 million, and that for a single investigator program! Research grants in that period were typically long-lasting, extending over the work of multiple graduate students. The AFOSR contract extended well over a decade, from 1964 to the late 1970s.

William Arthur Tiller was born in Toronto, Canada, on September 18, 1929. He grew up in Toronto, graduating from high school there in 1948. He studied engineering physics at the University of Toronto, receiving his B.S. degree (1952), M.S. degree (1953), and Ph.D. (1955). It was during his time as a graduate student that he began his work on solidification science. He did his M.S. thesis under the direction of Bruce Chalmers, a distinguished authority in materials science. Chalmers, later a professor at Harvard, was the founding editor of *Acta Metallurgica,* and was elected to the National Academy of Sciences in 1975. (Although his early work with Chalmers led to his Ph.D. dissertation, Tiller's dissertation was signed by his department head, since Chalmers had already left for Harvard by the time of Tiller's graduation.)

Tiller's M.S. thesis with Chalmers was published in 1953 in the first volume of the new journal *Acta Metallurgica*. It described the redistribution of solute atoms during the solidification of alloys and provided the first quantitative treatment of what became known as "constitutional supercooling." It became a classic paper in the field and by 2019 had been cited more than 1,300 times, making it Tiller's most frequently cited paper. Tiller's paper was co-authored with Chalmers, Kenneth Jackson and J.W. Rutter, who were working with Chalmers at the time. Both Jackson and Rutter went on the have distinguished careers of their own in solidification science.

As Tiller explained in his paper, when an alloy solidifies, the solute atoms that are rejected from the growing crystal are not uniformly distributed in the surrounding melt, but instead are concentrated near the crystal-melt interface because the diffusion of the solute in the melt is limited. The high concentration of solute near the interface determines the local liquidus temperature, which, in turn, determines whether the liquid is stable or unstable with respect to crystallization at that point. Instabilities then control the morphology of the interface between the growing crystal and the melt phase. Tiller's quantitative treatment of constitutional supercooling, which had been described qualitatively before by Rutter and Chalmers, is the treatment now found in textbooks covering solidification science.

Tiller's first two doctoral students, Kotler and Tarshis, both worked on extending Tiller's classic treatment of constitutional supercooling, by taking account of the kinetics of atomic attachment to the growing crystal at the crystal-melt interface. Their dissertations, completed in 1967, were the first in a series from the Tiller group that involved ever more complex refinements of the basic approach to solidification that Tiller had developed in his work with Chalmers.

In 1972, Tiller, with Ph.D. student Robert J. Asaro, published a paper in *Metallurgical Transactions* that was to become one of the most influential papers to come out of the department. While the paper was nominally on stress corrosion cracking, the theoretical ideas put forth were to have much broader and more far-reaching consequences. Asaro and Tiller pointed out, for the first time, that a flat free surface is unstable with respect to shape undulations if the solid is subjected to strain. Thus, if an initially flat surface of a strained crystal is slightly perturbed into a wavy shape, then atoms on the surface will spontaneously diffuse from the valleys to the peaks, causing the undulation to grow, eventually developing into sharp cracks at the surface. Asaro and Tiller suggested that this might be a mechanism for stress corrosion cracking. But much later, when strained epitaxial films of semiconductors were being used in microelectronic applications, the phenomenon was rediscovered and the Asaro-Tiller paper became the foundation for a new line of theoretical work on shape evolution of strained surfaces. By the end of the century, the Asaro-Tiller mechanism was being used to fabricate a wide variety of semiconductor quantum nanostructures, including quantum dots. What started as a simple idea to understand a corrosive failure process ended up being a conceptual tool for fabricating optoelectronic devices.

Tiller was best known for his theoretical and experimental investigations into the detailed physics of the freezing process for a wide range of materials, including water, metals, semiconductors, oxides and polymers. He investigated the relationships between the crystallization process and the resulting structures and properties. He exploited this knowledge of crystallization to generate new processes for ingot casting, material purification, and single crystal growth, among other technical problems.

In 1966, Shepard stepped down as executive head of the Materials Science Department, and Tiller took his place. Tiller must have already been discussing with Dean Joseph Pettit changes he wanted to make because, by the time he took office, a new senior professor, G. Marshall Pound, from Carnegie Tech, had been appointed and was already at Stanford. Pound's appointment would have been through a target-of-opportunity appointment, wherein the appointment is made without a formal search on the grounds that the candidate is so distinguished that he would certainly win any search that might be conducted. Tiller had known Pound since the 1950s when they both were in Pittsburgh, and both had contributed papers to the 1957 ASM symposium on Liquid Metals and Solidification.

As soon as Tiller took office that September, he talked about the many other changes he wanted to make. In his first meeting with the faculty, he urged the appointment of Arthur I. Bienenstock, then at Harvard, to strengthen the department in the area of structures and diffraction. Tiller had received a letter recommending Bienenstock for an appointment as an associate professor from David Turnbull, then a distinguished professor of Applied Physics at Harvard and an authority on glass formation in metals. Tiller knew Turnbull from his time at the GE Research Laboratory, where he had done seminal work on crystal nucleation during solidification of liquid metals.

Tiller was also thinking about ways to improve the visibility and financial status of the department. In that first faculty meeting, he urged the faculty to consider establishing an industrial affiliates program, as other engineering departments at Stanford had done. This would be studied and discussed during the fall quarter, and was completed by the early part of 1967. The industrial affiliates program was designed to establish closer connections between the department and the industrial organizations served by the department. Through periodic visits by faculty to the participating companies, and through an annual meeting of the affiliates, where they learned more about the work of the department, the companies were able to get an early look at research results before they were published and could get to know graduate students as prospective employees before they had graduated. Equally important was the opportunity to learn more about the long-range needs of industry and the contributions the department might make to satisfy those needs. The industrial affiliates were charged an annual fee, initially $5K, to participate in the program and were encouraged to provide funding to faculty members for projects of particular interest. The annual fees collected allowed the department to develop a much-needed unrestricted fund for the operation of the department.

The new department head also suggested discontinuation of the Master's Research Report as a requirement for the M.S. degree, revamping of the department's written and oral qualifying exams, and a change in the university oral examination structure. Tiller recommended that students taking their orals be given the questions in written form just prior to the exam, so that they could study and consider their answers before facing the examiners. This would be unfair to international students, he argued, who were not used to the verbal give and take of an oral exam in English. That practice was adopted and used by the department for about 30 years. Tiller also recommended that the department's university oral examination be patterned after that used by the Physics Department, where the candidate makes a public presentation of his/her thesis and is then questioned in a closed session following the presentation. Prior to this, students were simply questioned by a committee about the work they had done for their thesis and the contributions they had made and examined more broadly to determine their suitability for receiving a Ph.D. Tiller's recommendations were adopted and are still used to this day.

Tiller soon began thinking about how he would evaluate faculty performance and justify to the dean his recommendations regarding faculty salaries. He asked all faculty members to keep track of time spent on various activities such as teaching, administration, research, and advising, down to 15-minute intervals – timecards! He hoped that such data would enable him to better argue for higher faculty salaries. Although most of the faculty complied with the request, they were certainly not thrilled with the task. Pound, who had just arrived from Carnegie Tech, was especially cool to the idea. It was reported that his timecard simply read: "scholarly work—40 hours per week."

Another innovation that Tiller introduced was the use of small, hand-held tape recorders by faculty. In the days before personal computers, many faculty members would write out papers and communications by hand and then have them converted into typewritten documents by secretaries. Tiller bought tape recorders for each faculty member and gave each secretary a special listening device that would facilitate transcribing. The system was used for several years by some of the faculty. It was especially convenient for Pound who, by the late 1970s, had developed Parkinson's disease and found it increasingly difficult to write by hand.

The faculty minutes of 1966 include an interesting account of an effort by the department's graduate students to conduct their own independent evaluation of the department's professors, the courses they taught, and the facilities they had available to conduct research. The students proposed a detailed three-page evaluation questionnaire that included such provocative phrasing as whether a course was "shockingly insufficient," or that an instructor's preparation was "exceedingly poor," or that whether the student-instructor communication in a given course was "deplorable." Although Tiller thought the questionnaire could "serve a constructive purpose," the minutes report, the proposal was "strongly disapproved of by the faculty and was not set in motion." The faculty claimed that the well-established Tau Beta Pi surveys of university courses and instructors was sufficient. The affair, however, reflects the growing student activism of the time, though in quite a mild form compared to events elsewhere on campus.

William A. Tiller ca. 2000.
(Stanford University Archives)

Tiller continued his teaching and research while serving as department head in the late 1960s. In addition, he developed a growing avocational interest in psychoenergetics. His interest in this subject developed even further during his sabbatical leave in 1970-71, when he spent a year with his family at Oxford University with support of a Guggenheim Fellowship. It was during this period that he decided to give up the chairmanship of the department, as well as most of his committee work, so that he could pursue psychoenergetic research in parallel with his conventional science research at Stanford. As Tiller has written on a personal website about his work in psychoenergetics: "He was substantially supported in this by his wife, Jean, but substantially opposed by most of his Stanford Colleagues." John Shyne, who had been acting chair of the department during Tiller's sabbatical leave, became the chair in 1971. (By the end of the 1960s, the term Executive Head had been dropped in favor of the title Department Chair.)

Tiller continued his teaching and research in materials science during the 1970s and 1980s until his retirement in 1992. His work on the science of crystallization culminated in two books that he wrote in the early 1990s: *The Science of Crystallization: Microscopic Interfacial Phenomena* (1991) and *The Science of Crystallization: Macroscopic Phenomena and Defect Generation* (1992). In addition, Tiller has written six books on psychoenergetics and related subjects.

Richard H. Bube ~ Electronic Properties of Materials

The early 1960s had seen the initiation of teaching and research on electronic properties of materials. Richard H. Bube was appointed as an associate professor in 1962 (jointly with the Department of Electrical Engineering) and Robert L. White and William E. Spicer were appointed the following year as associate professors in electrical engineering, with joint appointments in the Materials Science Department. Together they were to provide teaching and research activities for several decades on electrical (Bube), photoelectronic (Bube), optical (Spicer), and magnetic (White) properties of materials.

Richard Howard Bube came to the department from the RCA Laboratories in Princeton, New Jersey, where he had been a senior member of the technical staff and a group director for research in photoelectronic materials. He had already published his first technical book, *Photoconductivity of Solids,* with Wiley in 1960. His first courses were on "Quantum Theory in Solid State Phenomena" and "Electronic Processes in Insulators." During his 30 years on the faculty, he trained 50 Ph.D. students and served as department chair for more than a decade.

Bube was born on August 10, 1927 in Providence, Rhode Island. He grew up in Providence and later did his undergraduate work at Brown University. He must have been a prodigy, as he was just 18 years old when he was awarded his Sc.B in physics from Brown, in 1946. He went on to Princeton University for graduate work in physics, receiving his M.A. degree in 1948 and Ph.D. in 1950, the latter at the age of 22. After completing his M.A., he joined the technical staff of the Radio Corporation of America (RCA) in Princeton, continuing his doctoral work at Princeton University with support of a contract jointly funded by the Office of Naval Research and RCA. The thrust of his initial work involved the study of luminescence in materials based on ZnS (zinc-sulphide); these materials were being developed as phosphors for television screens.

Richard H. Bube, 1962.
(Stanford News Service)

Richard Bube (left) and his visiting scholar, Guy Marlor (right), enjoying a chalkboard cartoon poking fun at them in 1966. (David Barnett)

During his 14 years at RCA, Bube published 60 papers, 34 of which were single-authored papers. He was such a prolific writer that a rumor developed among his Stanford students that he had a typewriter in the laboratory so that he could write papers as soon as the data came in. Through his skill as a writer, he helped many of his students learn to write well.

At Stanford, Bube built a research group on photoconductivity of solids with initial support from the U.S. Army Research Office (Durham) and the Center for Materials Research at Stanford. His first publication with a graduate student came in 1965 when he published a paper with Arthur L. Robinson on the effects of oxygen adsorption on photoconductivity of sintered layers of CdS-CdSe (cadmium sulfide - cadmium selenide) and CdSe (cadmium selenide). Robinson was to become one of Bube's early Ph.D. graduates, and went on to a very successful career in scientific writing, having served as a writer for *Science* for several years and as a writer for both the Stanford Synchrotron Radiation Laboratory at SLAC and the Lawrence Berkeley National Laboratory. He also wrote for the *MRS Bulletin* of the Materials Research Society.

Another one of Bube's most prominent students was Gerald B. Stringfellow, who received his Ph.D., in 1967, for work on the photoelectronic properties of ZnSe (Zinc-Selenide). Stringfellow joined Hewlett-Packard Laboratories following his graduation and quickly rose to become a project manager there. After working at HP Labs for 13 years, he accepted an appointment as Professor of Materials Science and Engineering and Electrical Engineering at the University of Utah. He was later made a Distinguished Professor there and served as Dean of Engineering for five years at the turn of the millennium. He became internationally known for his work on the growth and understanding of III-V semiconductor compounds, such as gallium-indium-nitride, which are important in the areas of fiber-optic communications systems and solar cells. He is also considered a pioneer for his work on light-emitting diodes as energy-efficient and long-lasting light sources. His C.V. is filled with awards and honors for this work, which was topped off by his election in 2001 to the National Academy of Engineering, one of only seven graduates from the department to have achieved that distinction to date.

Stringfellow reports that he had been interested in semiconductors since his teenage years, when he used GE transistors to build a radio. It was then a natural fit for him to work on semiconductors with Bube, since he was interested in those materials for solar cell and other applications. Stringfellow's project on ZnSe (Zinc-Selenide) for blue LEDs was a little outside of Bube's area but was a salutary choice for his future career. He loved the freedom and independence it gave him to pursue his own direction. Stringfellow has also reported that Bube was a very supportive and thoughtful advisor, arranging a summer job for him at the RCA Sarnoff Labs, where Bube had worked, and never failing to invite all his students to his home at Christmas.

The cartoon shown in the photograph on the previous page was drawn by David M. Barnett, at the time a graduate student and later a professor in the department. Barnett had been studying late at night with Art Robinson, one of Richard Bube's graduate students, when the cartoon was drawn. The drawing was inspired by the Batman TV series of the 1960s and depicted the frequent vigorous discussions between Bube and visiting scholar Guy Marlor. The next morning, Robinson appeared in Barnett's office quite red-faced with the report that he had forgotten to erase the chalkboard when he left for the night, as Barnett had asked him to do. As luck would have it, Bube had seen the cartoon in the morning, showed it to Marlor, and had this picture taken. Barnett sheepishly but bravely walked into Bube's office to confess and apologize, only to see Bube break out with a broad smile and laugh. Marlor had a wry Englishman's sense of humor, and he, too, enjoyed the cartoon as it illustrated the warm relations between graduate students and faculty in the department.

Bube's first technical book was published before his coming to Stanford. It was the first of what would become a very long list of textbooks written during his career. Counting the different editions as separate books, Bube had written ten textbooks by the end of his Stanford career, including *Electronic Properties of Crystalline Solids: An Introduction to Fundamentals* (1974), a second edition of *Photoconductivity of Solids* (1978), *Electrons in Solids: An Introductory Survey* (1981, 1988, 1992), *Fundamentals of Solar Cells: Photovoltaic Solar Energy Conversion* (1983, with A.L Fahrenbruch), *Photoelectronic Properties of Semiconductors* (1992), *Photo-induced Defects in Semiconductors* (1996, with David Redfield), and *Photovoltaic materials* (1998).

Richard Bube was a deeply committed Christian who lectured and wrote about science and Christianity throughout his adult life. He taught at the Fuller Theological Seminary, a multi-campus evangelical seminary with a campus in nearby Menlo Park, and Regent College (Vancouver, British Columbia) and was frequently a Staley Distinguished Christian Scholar Lecturer, an annual lecture series aimed at bringing distinguished evangelical speakers to campuses nationwide. He also served

Richard H. Bube, 1978.
(Stanford News Service)

as elder, teacher, or lay preacher in local churches affiliated with Lutheran, Presbyterian, and Covenant denominations. For more than 20 years, he conducted an undergraduate seminar at Stanford on "Issues in Science and Christianity," until it was cancelled in 1988. He wrote five books on Christianity and Science including: *To Every Man an Answer: a Systematic Study of the Scriptural Basis of Christian Doctrine* (1955), *The Encounter Between Christianity and Science* (1968), *The Human Quest: a New Look at Science and the Christian Faith* (1971, 1976), and *Putting it all Together: Seven Patterns for Relating Science and the Christian Faith* (1995). He also wrote his 530-page autobiography entitled *One Whole Life* (1995). Taken together, he published sixteen books, making him the most prolific author to have served on the Materials Science faculty thus far.

While Bube continued to have a very active teaching and research program in the 1970s and 1980s, he was called on to be department chair in 1975 when John Shyne stepped down. Bube was considered a good choice in part because he was so efficient that he could handle the job of chair while continuing his research and teaching.

Bube's term as chair was characterized by stability; he "kept the trains running on time." After a flurry of faculty appointments in the 1960s, the faculty had reached a stable size of about 12, which remained essentially constant from the early 1970s to the early 1990s. Bube tried to foster a family atmosphere in the department that was reflected in the picnics and Christmas parties that he promoted and enjoyed. He also started an annual department *Newsletter* that told of the comings and goings of people in the department and highlighted department events of broad interest. The newsletter was published and printed annually for most of the period from 1975 to 2007, until digital information displaced that form of hard-copy communication.

Bube retired in 1992 to devote more time to his writing and speaking on science and faith. He also was able to spend more time with his family and friends and with his church. He died on June 9, 2018 and, as his son said, "He went home to his Lord."

William D. Nix ~ Building the Junior Faculty Ranks

In the early 1960s, just as Stanford also won an ARPA contract for the Center for Materials Research, the School of Engineering secured a major grant from the Ford Foundation for faculty development. With that support, the Materials Science Department continued building up its faculty. In 1962, William (Bill) D. Nix, then finishing his Ph.D. degree in the department, was appointed as an acting assistant professor, in anticipation of his regular appointment as assistant professor in 1963.

Nix was already well-known for the quality of his teaching, having taught part-time at San Jose State College throughout his Stanford

William D. Nix, 1964. (W.D. Nix)

Norman Ahlquist and William Nix admiring a newly acquired electro-hydraulic mechanical testing system, one of the first to be sold by the Materials Testing Systems (MTS) company. (Stanford University Archives)

graduate studies. He was promoted to associate professor (1966), and professor (1972). He would serve on the regular faculty for 40 years, including five years as department chair, before becoming professor emeritus in 2003.

During his time on the faculty, he co-authored two textbooks, published hundreds of papers with his Ph.D. students, and earned many awards for both his teaching and his research, including the top teaching awards from both ASM International and The Metallurgical Society, and the Von Hippel award from the Materials Research Society. Later in his career, he was elected to the National Academy of Engineering, the American Academy of Arts and Sciences, and the National Academy of Sciences. In 2019 TMS, The Minerals, Metals and Materials Society, established The William D. Nix Award, an annual award and lectureship to be given each year at the annual meeting of that society. It follows two other lectureships of that kind created in the 1970s named for Robert Franklin Mehl and William Hume Rothery.

William Dale Nix was born on October 28, 1936 in King City, a farming and ranching community in the Salinas Valley of California's central coast. His parents had arrived there from Arkansas in 1933 as a part of the dust bowl migration that John Steinbeck chronicled in 1939 in his Pulitzer Prize winning novel, *The Grapes of Wrath.* After living in the King City area, and then, during the Second World War, in Paso Robles near Camp Roberts, he moved with his family to San Jose. He finished his elementary education in San Jose, graduating from San Jose High School in 1954, some 35 years after O. Cutler Shepard's 1919 graduation from the same school.

In high school, Nix became interested in drawing and thought he wanted to study architecture in college. But his mediocre high school academic record limited his options. He enrolled at San Jose State College with a plan to major in pre-architecture, hoping to transfer later to a university with an architecture program. His continuing poor academic work, however, together with his impression that there seemed to be too much "art" and not enough "tech" in architecture, caused him to abandon that plan altogether. He elected to major in engineering, thinking it a better path to a viable job, especially if he were to flunk out of college, which, he admits, he thought likely. As it turned out, engineering did indeed turn out to be the happier and more productive path.

Early in 1957, during his third year, he saw an announcement for an ASM scholarship for a student who wished to study metallurgical engineering. Liking the subject and needing the $400 that the scholarship would bring, he applied, selecting metallurgical engineering as a major. Awarded the scholarship later that year, he was on his way to a materials career. He finished his B.S. degree in metallurgical engineering, with much improved grades, in 1959, and decided earn a Ph.D. degree so that he could teach at the college level. He was admitted to the Metallurgical Engineering Department at Stanford on January 20, 1959.

Nix had followed what may be called an unlikely path to a career in academics. As a boy, he was so unimpressive that a science teacher told his mother that while he was a nice boy, "he should not look for a very challenging line of work." His parents wanted him to have a good education, however, something they themselves had been denied in the economic downturn (his father had finished 7th grade and his mother the 10th grade) and he would be the first in his family to go to college.

While during his first two years at San Jose State his only A's were in basketball, it all

William D. Nix, 1985. (W.D. Nix)

changed with a basic course in statics, taught by Roy Zimmerman, a practicing civil engineer and part-time teacher in engineering. Zimmerman ran the course a little differently than other teachers, with frequent quizzes rather than exams. Nix discovered, to his great surprise, that he was not only doing as well as others, but better, and received an A in the course, the first of many to come. Nix also found his passion for teaching: He was so excited about his studies that he began to teach other students who still were struggling.

Due to his "late development," his undergraduate transcript looks rather strange today, starting off with C's, D's and F's and ending with A's. In the late 1990s, as the Materials Science faculty was sorting through applicants to its graduate program, Nix thought it would be fun to see how his own academic record would stack up with the applicants of the day. He re-typed his transcript, removed his name and the institution where his degree was granted, and posted the record outside his office with the question: "Should we admit this guy to our graduate program?" He left a sheet of paper for people to respond and make comments. Many graduate students, even some faculty, were hard on this mystery applicant; the comments were uniformly negative and pessimistic about the applicant's prospects, pointing to the very poor marks in math and engineering in the first two years but ignoring the high marks of the last two years. Needless to say, there was some embarrassment, and a little chagrin, when Nix revealed that the academic record displayed was his own.

Nix supported his graduate studies at Stanford by serving part-time as an instructor and acting assistant professor at San Jose State College. Since he was not tethered to a specific research grant, he was free to choose a dissertation subject of his interest, with the agreement of his advisor, Robert A. Huggins. Huggins had given a seminar course on magnetic properties of materials that Nix found interesting. He wrote a dissertation in that general area even though he later had become more interested in dislocations and mechanical properties of materials.

As a faculty member, Nix was mainly interested in problems relating to high temperature creep and fracture of heat resisting metals and alloys of importance in gas turbine engines and nuclear reactors. From the early 1960s to the mid-1980s, most of his work focused on the mechanisms of creep, diffusional deformation, superplasticity, dispersion

O.D. Sherby and W.D. Nix and their students and former students at the Sixth International Conference on Creep and Fracture of Engineering Materials and Structures at U.C. Irvine in 1997. Top row, left to right: Paul Kitabjian, Mike Uchic, Kevin Hemker, Jeff Wadsworth, Mike Kassner, Mike Mills, Oscar Ruano, John Stephens. Bottom row, left to right: T.G. Nieh, Jim Earthman, Erik Taleff, Oleg Sherby, William Nix, Glenn Daehn. (W.D. Nix)

strengthened metals, and intergranular fracture. By the mid-1980s, however, it became clear to him that a number of mechanical behavior problems were arising in microelectronics and other thin film technologies that needed attention. His former colleague, Craig R. Barrett, who after leaving Stanford became a rising star at Intel, had written to the department to urge his colleagues to move away from traditional materials to the microelectronic materials of interest to local industry. Another factor motivating the change in direction was that most of the department's graduating students were being employed in microelectronics. Nix and David M. Barnett were also consulting with Intel about problems in this area, but methods for studying mechanical behavior at the sub-micrometer scale were not broadly available to the department at that time, and very little attention had been given to understanding mechanical properties of materials at that scale. A major effort was needed to address problems of interest in microelectronics so Nix and others then began to turn their attention to problems of the mechanical behavior of thin films and other small-scale structures.

In 1983, Nix became aware of the development of the first commercially successful nanoindenter by his former student, Warren Oliver, and his colleague John Pethica (now Sir John Pethica). Nix realized that this instrument could be very useful in studying the mechanical properties of thin film materials. He acquired the first such instrument from Nano Instruments Inc. and began using it to study the mechanical properties of thin films. With his student Mary Doerner, he developed the first broadly applicable method for analyzing indentation experiments and extracting fundamental mechanical properties such as hardness and elastic modulus. A greatly improved analysis method was published six years later by Warren Oliver and George Pharr, who also had been one of Nix's students. The Oliver-Pharr paper established the methodology of nanoindentation that is now accepted worldwide. That 1992 paper became, arguably, the most highly cited paper in the field of materials science.

For their work in developing nano-indentation as a broadly useful technique for studying the mechanical properties of materials at small dimensions, both George Pharr and Warren Oliver were elected to the National Academy of Engineering, in 2014 and 2016, respectively, making them the second and third of Nix's students to be elected to NAE. The first was David K. Matlock of the Colorado School of Mines, who was elected to NAE in 2003 for his work on steel research and for the creation of a major steel research center at the Colorado School of Mines.

Nanoindentation experiments on metals showed that hardness at the sub-micrometer scale is not independent of depth, as it is at the bulk scale. Instead, an indentation size effect characterized by "smaller is stronger" is observed. The indentation size effect, which is associated with gradients of strain, attracted

William D. Nix, 2015. (W.D. Nix)

considerable attention in both the materials and mechanics communities.

Nix and his students also found another "smaller is stronger" effect using a micropillar compression technique that one of his former students, Michael D. Uchic, had developed in his work at the Air Force Research Laboratory. The surprising finding was that there is also a very strong "smaller is stronger" size effect in spite of the absence of strain gradients. This finding led to a major effort to understand this kind of size effect using the theory of dislocations.

Another Nix student, Julia R. Greer, a professor at Caltech, made use of the "smaller is stronger" effect to create ultralightweight materials called structural meta-materials. These materials have attracted much attention and are expected to have a wide variety of applications in biomedical devices, ultralightweight batteries, and damage-tolerant cellular solids.

In addition to his teaching and research, Nix took his turn in various administrative capacities in the department. He served for two years as Director of the Center for Materials Research at Stanford when Robert A. Huggins, the permanent director, was in Washington running ARPA's Interdisciplinary Laboratory Program in Materials. When Richard Bube became department chair in 1975, Nix became his associate chair and held that post until 1986. He was appointed chair of the department in 1991, following the term of Stig Hagström, and served in that capacity for five years.

After his stint as department chair, Nix happily returned to teaching and research. Except for his four-year service on Stanford's Advisory Board (the seven-member faculty board that makes final recommendations to the university's president on faculty appointments, promotions, and tenure), he devoted most of his time to his students and former students. By the end of his career, Nix had graduated 79 Ph.D. students, many who had gone on to academic careers of their own. Nix used to joke that he spent much of his time in those years making good on the warrantees that he had given with each graduated student. He became Professor Emeritus in 2003 but continued to teach and do research with doctoral students for another decade. In 2016, he and his colleague Wei Cai, in Stanford's Mechanical Engineering Department, published *Imperfections in Crystalline Solids*, an intermediate level textbook based on the graduate course, MSE 206, "Imperfections in Crystalline Solids," that Nix had given in the department for decades and that had been taken over by Cai and moved to Mechanical Engineering. The book was published by Cambridge University Press as the first in a new series of textbooks on materials fundamentals.

Chapter 8

Appointments in Metallurgy and Failure Analysis

John C. Shyne was appointed to the faculty in 1960, the first appointment to be made after the 1958 expansion of the faculty from two to five. Shyne was a physical metallurgist with an interest in phase transformations and mechanical properties of metals. After coming to Stanford, he developed an interest in consulting for law firms in the valley on the causes of structural failures and was joined in that effort by Alan S. Tetelman and G. Marshall Pound when they arrived on the faculty a few years later. These three, together with two additional partners, founded the company Failure Analysis Associates, Inc. (later, Exponent Inc.), without question the most important company to be created by people from the Materials Science department.

John C. Shyne ~ Phase Transformations, Internal Friction and Failure Analysis

John (Jack) Shyne came to the department after several years at the Ford Scientific Laboratory in Dearborn, Michigan, and a brief stint at the Mueller Brass Company in Port Huron, Michigan. He is a Michigander through and through, having been born in Detroit in 1925, and having lived in Michigan until he came to California. When he was 10, he moved to nearby Belleville, Michigan, where his mother's family was living, and soon thereafter, in the fourth grade, met his wife to be, Shirley Ann. After graduating from high school in 1944, Shyne was immediately drafted into the army to serve in a field artillery unit. In early 1945, after training at various camps in Arkansas and North Carolina, Shyne sailed to Europe on the Queen Elizabeth to serve in France as a cannoneer in the 80th infantry division. Later that year, as the war with Germany was ending, he was sent back to the U.S. to prepare for the planned invasion of Japan. The Japanese surrender in August of 1945 ended that plan, and he spent the remainder of his service time in the U.S.

By the fall of 1946, Shyne was married and back living in Belleville, supported by the G.I. bill as a member of the 52-20 club (a plan providing returning service men with $20 per week for 52 weeks). He enrolled at the University of Michigan, Ann Arbor, only 18 miles from his home. Following the path of an admired uncle, a metallurgist, he elected to major in both mathematics and metallurgical engineering. He received degrees in those subjects in 1951, and a year later received an M.S. degree in metallurgical engineering. In 1952, Shyne joined the Ford Scientific laboratory at Henry Ford's famous Rouge River Plant in Dearborn to work with Donald N. Frey, a group leader there who had been a faculty member at

John C. Shyne, 1960. (Stanford News Service)

Michigan. There, he worked with a colleague, Eric Morgan, to improve the formability of low carbon steels by using boron additions to remove nitrogen, which was known to be the cause of strain aging and inhomogeneous plastic deformation.

While working at Ford, Shyne took advantage of a program that permitted Ford engineers to continue working while pursuing advanced degrees at the University of Michigan. He arranged a Ph.D. project with Professor Maurice J. Sinnott that allowed him to do much of his research on the job. The result was his 1958 doctoral dissertation on iron-aluminum alloys using internal friction, a technique that he would later bring to Stanford.

Shyne started to take on consulting jobs involving failure analysis soon after he arrived at Stanford. Some of these jobs were passed on to him by O. Cutler Shepard, who was then too busy for consulting, given his duties as executive head of a growing department. As most of these consulting jobs involved fracture in one way or the other, Shyne brought in Alan S. Tetelman, an expert in fracture. Tetelman (about whom much more is written below) had been appointed in 1964. The collaboration between Shyne and Tetelman is regarded as a precursor of the founding, a few years later, of Failure Analysis Associates, which became one of the largest and most successful failure analysis companies in the world.

Shyne spent a sabbatical year (1969-1970) with Tarik Ogurtani (one of Huggins' former students) at the Middle East Technical University in Ankara, Turkey. Following that leave, he became acting chair of the Materials Science Department, filling in for William A. Tiller, chair since 1968. When Tiller stepped down from the chairmanship to devote more time to other interests, Shyne succeeded him. Shyne's period as department chair, from 1971 to 1975, was one of relative stability for the department, even if the rest of the country was still in a turbulent mood. It was during Shyne's term as chair that the department's name added "and Engineering," a change that Shyne and Sherby, among others, had been pushing for since the 1960s.

Shyne has reported that during his time as chair, the department tried to attract John Hirth of Ohio State University to the faculty, but the response from the new Dean of Engineering, William M. Kays, was an emphatic NO! Kays, who would serve as dean from 1971 to 1984, put great stock in attracting undergraduate students to a discipline, and he was not inclined to add faculty to a department that did not attract undergraduates in significant numbers. While Material Science faculty had built an impressive graduate program, and was developing a national reputation in research, the department's undergraduate program had withered and was barely alive. The relative neglect of the undergraduate program continued until the 1990s, when a new set of faculty members pushed for rebuilding the undergraduate curriculum.

Following his stint as chair, Shyne accepted a two-year appointment as section head of metallurgy and materials in the Division of Materials Research at the National Science Foundation. In that post, he supervised programs supporting research in metallurgy, ceramics, and polymers. He returned to Stanford after his service in Washington, D.C. in a most unusual way. He biked home! While not known as an athlete, Shyne did do some biking while in Washington, and he decided to ride his bike all the way back to Palo Alto. A party of faculty colleagues and friends were on hand to meet him on the day he was scheduled to arrive. Huggins remembers riding out on his own bike to see if he could meet Shyne as he was approaching home but gave up his search after a while, having seen only a very thin man of 51 with a floppy hat on a bicycle, whom he did not recognize. That was, of course, John Shyne, who had lost considerable weight on his several weeks' journey back to California.

Throughout most of Shyne's time at Stanford, he consulted with Failure Analysis Associates, Inc. (later, Exponent Inc.), the Menlo Park company that he and his colleagues had founded in the 1960s. He retired from Stanford in 1986 to work full time at Exponent and retired from Exponent ten years later.

Alan S. Tetelman ~ A Meteoric Career

The Merriam-Webster dictionary gives the following as a definition of meteoric: resembling a meteor in speed or in sudden and temporary brilliance—as in, a meteoric rise to fame. Most of this applies to Alan S. Tetelman who was appointed to the faculty at the age of 28 and died just 14 years later. Tetelman already had established himself as an authority in fracture of materials by the time of his appointment as an associate professor in 1964. He quickly established a vibrant research group, and by age 31 had co-authored an authoritative textbook entitled *Fracture of Structural Materials*. The 31-year old also co-founded the consulting firm Failure Analysis Associates, spent a year as deputy director of the Materials Science Program in the Department of Defense's Advanced Research Projects Agency (ARPA) and, a year later, accepted appointment as chairman of the Materials Department at UCLA. His career continued to shine brilliantly until he lost his life, at 42, in the tragic air crash over San Diego on September 25, 1978.

Tetelman's life and career were beautifully described in an obituary written in 1980 by his UCLA colleagues, Kanji Ono, Christian N.J. Wagner, Russell A. Westmann, and Alan Ardell:

> Alan Stephen Tetelman was born in New York on May 9, 1936. He attended the Riverdale Country School in Bronx, NY and after graduating from high school there he attended Yale University where he obtained his B.S., M.S., and Sc.D. degrees, the latter in 1961, all in metallurgy. Always a multi-dimensional individual, he played football for the Yale freshman team, fostered a keen interest in politics, and spent several summers working in the Texas oil fields. His Sc.D. dissertation on "The Mechanism of Hydrogen Embrittlement of Iron Alloy Single Crystals," was an experimental and theoretical tour de force, and immediately established him as one of the young potential

Alan S. Tetelman, 1964. (Stanford News Service)

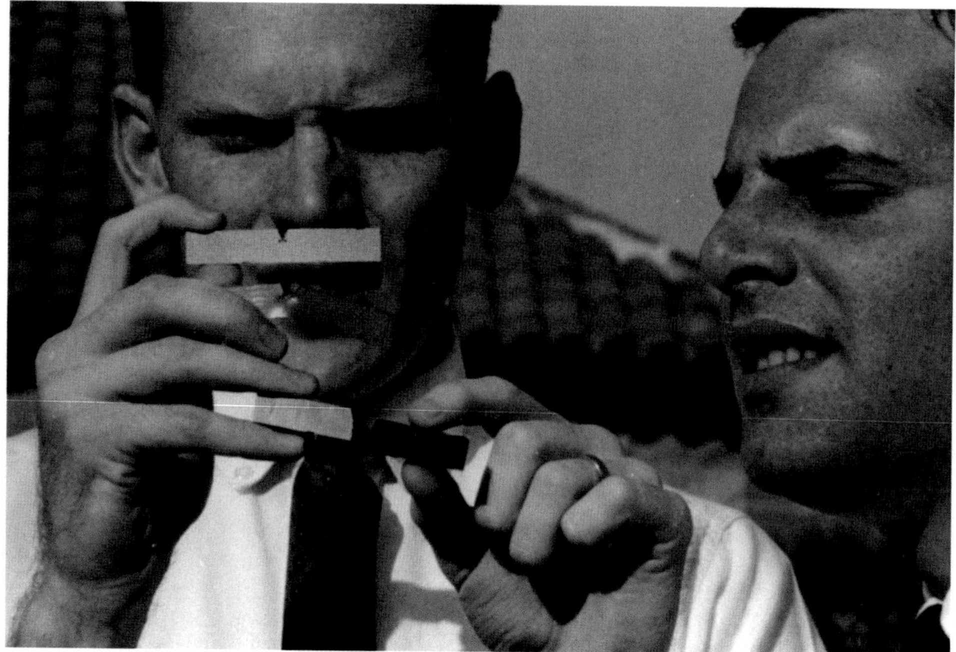

Charlie Rau and Alan Tetelman inspecting broken Charpy bars from their fracture experiments, 1960s. (Stanford University Archives)

superstars in his field. There was little doubt that Tetelman would live up to these early expectations.

On completion of his graduate work, he spent a year at the University of Paris as a National Science Foundation postdoctoral fellow where he worked with the famous Jacques Friedel. Tetelman was always a Francophile, spoke French with reasonable fluency, and took advantage of this postdoctoral period to hone a variety of skills, scientific and otherwise. He returned to the United States to join the Ford Motor Company Scientific Laboratory where he spent two years as a staff research scientist. He already had an international reputation when he left Ford to come to Stanford. *(Tetelman Memorial Resolution, 1980)*

In 1965, Tetelman was awarded the Robert L. Hardy Gold Metal of AIME, which is presented annually to the outstanding young metallurgist in the country under the age of thirty (Robert Huggins had received that award earlier (1957), and Craig Barrett (1967) and Robert Sinclair (1976) would win it a few years later).

A sense of Tetelman as a young man in a hurry can be judged from the following anecdotes. Soon after he joined the faculty, he invited William Nix, then an assistant professor, to join him on a walk around the Stanford campus to discuss their careers in the field. Nix remembers Tetelman saying emphatically that they both needed to "develop reps" as soon as possible, meaning develop their reputations, which, in fact, he already had done. But the conversation underscored the intensity with which he was to pursue his career.

At about that time, Tetelman and Nix were together at a conference of The Minerals, Metals and Materials Society (TMS) where some sort of mechanical behavior was being discussed. Someone had given a talk in one of the technical sessions and it was being discussed by some of the senior luminaries in the field, including the highly respected Dr. James C. M. Li of the U.S. Steel Fundamental Research Laboratory. Nix was stunned to hear Tetelman call Dr. Li "Jimmy" in the open discussion. It was another small announcement that Tetelman had arrived and was to be reckoned with.

Again, quoting extensively from Tetelman's UCLA memorial resolution:

Tetelman left Stanford in 1967 to spend a year on temporary assignment as deputy director of the Materials Science Program in the Department of Defense's Advanced Research Projects Agency (ARPA). While there, he recognized the importance of nondestructive evaluation for assuring the integrity of stress bearing structures. Among many approaches, he was most attracted to the potential of acoustic emission and was instrumental in the birth of two firms specializing in the manufacture and application of acoustic emission instrumentation. The two firms, Dunegan Research Corporation (later Dunegan/Endevco) and Acoustic Emission Technology Corporation, grew substantially, each with millions of dollars in annual sales. Tetelman also incorporated this new research tool for fracture mechanics studies, such as the fracture toughness and hydrogen embrittlement of materials, and failure modes of composites.

After his year at ARPA, he had intended to return to Stanford, but was offered the opportunity, at age 32, to assume the position as chairman of an incipient Materials Department at UCLA. This presented a challenge he could not resist, and in a veritable explosion of hiring activity he directed the growth of the Materials Department from a faculty of seven to one of seventeen within a time span of two years. He served the Materials Department as its chairman until 1974. At the time of his death, Tetelman was on leave from the University, to devote increasing effort to failure analysis activities in Southern California. He was on his way to San Diego to investigate an airplane crash when the mid-air collision that claimed his life occurred.

Alan Tetelman was a mercurial individual. Few people could think more quickly on their feet. He was decisive and unafraid to defend positions unpopular with friends, students, colleagues, and adversaries. He had an uncanny ability to convince himself that his particular course of action was absolutely proper but was nevertheless quick to recognize and adopt a superior alternative. Not everyone responds to such a person with magnanimity, but the respect which he gained, sometimes given grudgingly, was nearly universal.

G. Marshall Pound ~ Statistical Thermodynamics with Color

G. Marshall Pound (called "Marsh" by friends and colleagues) came to the department in 1966 as a full professor from Carnegie Institute of Technology, where he had been professor and director of the Metals Research Laboratory. He was an authority on kinetics, especially nucleation, having authored a number of papers and book chapters on the application of statistical mechanics and thermodynamics to the nucleation of solid and liquid phases. He had a reputation at Carnegie Tech as an outstanding classroom teacher, which was quickly confirmed by students and faculty at Stanford.

Many remember Pound's ability to give complex and intricate lectures on statistical thermodynamics without notes! He would meditate in his office before going to the classroom and write out all of the derivations on the chalkboard by memory. Others remember his ability to impersonate a moving ledge on a surface during evaporation or condensation. He would typically tip toe across the front of the lecture hall, looking furtively over his shoulder at the other ledges trying to catch up with him.

G. Marshall Pound, 1968. (Stanford News Service)

Marshall Pound at an MSE graduate student/post doc party in the 1970s. From left to right are: Jay Spingarn (Nix), Paul Gilman (Nix), George Pharr (Nix), Marsh Pound, David Bourell (Sherby), unknown (with glasses), Steve Goods (Nix), Jeff Wadsworth (Sherby), Paul Fuoss (Bienenstock), Rick White (Nix), Gary Michal (Sinclair) and Jer-Hong Lin (Sherby) (parentheses indicate academic advisors). (Paul Gilman)

Another favorite memory of students was his impersonation of a drunken sailor in connection with his description of the random walk problem in diffusion. He was a legendary teacher and mentor of graduate students who loved him as they would a father.

Guy Marshall Pound was born in Portland, Oregon, on April 2, 1920. His family moved to The Dalles, Oregon, where he received his secondary education. He enrolled in Reed College in 1937 and was granted a B.A. degree in chemistry in 1941. After receiving an M.S. degree in chemical engineering from MIT, in 1943, he worked for a while as a chemical engineer in California before enrolling in a doctoral program in physical chemistry at Columbia University, receiving his Ph.D. in 1949.

His career in metallurgy and materials science started in 1949 when he was hired by the legendary Robert Franklin Mehl of Carnegie Tech to become an assistant professor in the Metallurgical Engineering Department. At the time, the department at Carnegie Tech was arguably the top department in the country in the field of metallurgy. Pound's assignment was to introduce the disciplines of chemical thermodynamics and kinetics to the department. What followed, in the 17 years he was on the Carnegie Tech faculty, was a revolution in the way the subjects of thermodynamics and kinetics were taught in departments of metallurgy. He and his Carnegie Tech colleagues were responsible for developing the high-level science now routinely applied to these subjects in materials science curricula around the world. Many of the outstanding professors in this area of materials sciences studied with Pound, most notably John P. Hirth (mentioned above) who taught at Ohio State and Washington State Universities.

Pound's 30-year research career included many outstanding contributions to metallurgical knowledge, earning him much recognition for his work. He is best known for his works on kinetics of crystal growth from the vapor, statistical mechanics of nucleation theory, and computer simulation of phase transformations. His work on the kinetics of crystal growth culminated in a monograph entitled "Condensation and Evaporation: Nucleation and Growth Kinetics," published in 1963 with his former student, John Hirth, as a co-author. That work stood as a standard reference in the field for two decades. His most notable contribution to the theory of nucleation was made in 1966 in a paper co-authored by Jens Lothe. Their paper, entitled "On the Statistical Mechanics of Nucleation Theory," laid the foundation for what is now known as the Lothe-Pound Theory of Nucleation. In later years, Pound became interested in using the techniques of computer simulation to describe nucleation and growth of crystals. Much of his work in this area was done in collaboration with workers at the IBM Research Laboratory in San Jose. Through his research on these subjects, Pound trained 35 PhD students who continued the work he started.

As noted earlier, soon after Pound arrived at Stanford, he joined with Alan Tetelman and others to form Failure Analysis Inc., a consulting company specializing initially in the investigation of accidents. The company prospered and grew during the 1970s and 80s and became the largest company of its kind in the world. In connection with his work in failure analysis, Pound regularly involved graduate students and post-docs in the conduct of experimental investigations. In addition to paying them for their work, Pound would frequently invite the whole gang to his home in Palo Alto Hills. The photograph on the previous page shows Pound at a graduate student party at that time.

Pound received much recognition and many honors during his career in materials science. He held Senior Fulbright and Guggenheim Fellowships when he studied at the University of Sheffield in 1959-60, and a Guggenheim Fellowship and Visiting Professorship at the University of Berlin in 1964. He was a fellow of both ASM International (1979) and The Metallurgical Society of AIME (1983). In 1978, he received the Albert Easton White Distinguished Teaching Award of ASM, and in 1984 he was selected by The Metallurgical Society to deliver the Institute of Metals Lecture and to receive the Robert Franklin Mehl Award. His lecture, entitled "Perspective on Nucleation," was an appropriate culmination of his professional career and scholarly work in the field of Material Science.

In the late 1970s, Pound contracted Parkinson's disease, which ultimately shortened his career and forced him into retirement. His disease limited his ability to write on the chalkboard, though he continued in the last few years to be active in all other ways, especially using the small tape recorder that Tiller had provided in the late 1960s. He took medical retirement in 1980 and moved with his wife, Barbara, to Crescent City in Northern California, where he died in 1988.

Failure Analysis Associates

The story of the Materials Science Department would not be complete without giving an account of the founding of Failure Analysis Associates, surely the largest and most successful company ever to come out of the department. As noted above, John Shyne and Alan S. Tetelman started working together on failure analysis consulting jobs as soon as Tetelman joined the faculty in 1964. When G. Marshall Pound joined the faculty, in 1966, he, too, worked with them on consulting. By this time, Tetelman was taking the lead on these consulting activities and was accepting a wider set of assignments and working with a wider group of people, including engineers at Stanford Research Institute (SRI) in Menlo Park. By 1967, Tetelman and the others had decided to form a company to be called Failure Analysis Associates (FaAA), to, as they said, "pioneer a new field—investigating how and why failures and accidents occur." The primary inspiration for the founding of the company came from Tetelman. He and Bernard Ross, an engineer at SRI, were co-founders of the new company. Professors John C. Shyne and G. Marshall Pound, along with Sathya V. Hanagud, also an engineer at SRI, were original partners in the company. As an interesting aside, the founders and partners met to form the company at The Oasis, a beer and burger joint in Menlo Park. They each put up $100 to make the $500 capitalization for the company, the only capitalization the company has ever used.

In the early 1970s, after Tetelman had left Stanford to become chair of UCLA's Materials Department, Failure Analysis Associates was still a loose collection of professors and failure experts doing consulting work for various companies and organizations. Some of this work was coming from the Electric Power Research Institute (EPRI) in Palo Alto. Failure Analysis Associates had the inside track on this work, in part, because Tetelman was in close contact with Dr. Chauncy Starr, president and founder of EPRI. (Starr had hired Tetelman at UCLA a few years earlier when Starr was Dean of Engineering and Applied Sciences there.)

Alan S. Tetelman on a Failure Analysis job. (Exponent Inc.)

Tetelman soon realized that, in order to execute these contracts, he would have to change FaAA from a loose collection of professors and investigators into a real company. He recruited Charles A. Rau, then at Pratt and Whitney, to be the company's first full-time employee. Rau had been one of Tetelman's first Ph.D. students at Stanford, and he succeeded admirably in this task. His appointment was a major turning point for the firm, as he established a solid foundation for future company success.

As stated on the company website:

> FaAA initially provided engineering expertise to the legal and insurance industry. In the 1970s, the Company continued to expand its accident investigation services while also responding to the increased engineering requirements of the utility and nuclear power industries. In the 1980s, the Company added new engineering and science disciplines including civil engineering, structural engineering, electrical engineering, biomechanics, data analysis/statistics, human factors and visual communication services. We were well known and respected for our independent investigations on some of world's biggest engineering accidents and failures and we expanded our office footprint outside of California.
>
> During the 1990s the Company continued to grow and expand through hiring and acquisition. In August of 1990, the firm went public, trading on the NASDAQ exchange under the symbol FAIL. We soon realized that the firm offered a broader range of services outside of just "failure analysis" and needed to reflect that in our brand. In March 1998, we changed our name to Exponent (NASDAQ: EXPO) which means "one who expounds or interprets" - which better represents the services we offer to our clients.

Failure Analysis Associates (Exponent) has investigated some of the most prominent structural failures, disasters, and accidents of the past half-century, and has contributed enormously to the development of tools and methods for predicting and preventing such devastating events. The company investigated the collapse of the walkways spanning the atrium of the Hyatt Regency Hotel in Kansas City in 1981, where 114 dancers were killed in the collapse and 200 others injured. They found that a change in the floor hanger arrangement during construction, while not changing the stress in the rods holding up the walkways, doubled the loads at the floor connections and led to the collapse.

The company also investigated the bombing of the Alfred P. Murrah Federal Building in Oklahoma City in 1995, which killed 168 people and injured more than 750. They determined that the bomb consisted of nearly 500 pounds of ammonium nitrate fertilizer combined with fuel oil, which resulted in an energy equivalent of two tons of TNT. The bomb was in a truck parked in a loading zone only 10 feet from the building. The company report led the Federal Government to make changes in security procedures to strengthen government buildings against such attacks.

The company's investigation of the terrorist attack of the Twin Towers of New York City's World Trade Center on September 11, 2001 led to the understanding of why the buildings collapsed. The towers ultimately failed by plastic buckling of the steel columns. The columns were stripped of their fireproofing by the force of the impact and subsequently were heated by the fires started by the burning jet fuel to the point where they lost their strength and ability to support the floors above the site of the crash. Nearly 3,000 people were killed in the attack.

As of 2016, Exponent had become an international company with offices all over the world (20 in the United States, four in Europe and two in China), with almost 1,000 employees and a consulting staff of more than 700, and with revenues exceeding $300M. This is the company that Alan Tetelman and his colleagues founded at The Oasis in 1967 with $500 in capital funding.

Portrait of Alan S. Tetelman at Exponent, Inc. (Exponent Inc.)

Tragically, Alan Tetelman was killed on September 25, 1978, in the PSA Flight 182 air crash over San Diego between the jet liner and a private Cessna airplane, an accident that claimed the lives of 144 people. Ironically, he was traveling to San Diego to investigate the cause of a Navy jet crash. He is remembered as one of the most charismatic people ever to have worked in the department.

In the 50-year history of the company, many people who studied in the MSE department at Stanford joined Failure Analysis Associates (Exponent) and some reached leadership positions in the company. Among the doctoral students who went on to work at the company were (name, Ph.D. year, thesis advisor, and highest company position):

Charles A. Rau, Ph.D. 1967 (with Alan Tetelman), Group Vice President and Principal Engineer

Anton J. West, Jr., Ph.D. 1970 (with Craig Barrett), Hourly

Lee A. Swanger, Ph.D. 1972, (with G. Marshall Pound), VP and Principal Engineer

Thomas L. Read, Ph.D. 1972 (with William Nix), Hourly

Richard W. Lund. Ph.D. 1975 (with William Nix), Manager

Steven H. Goods, Ph.D. 1977 (with William Nix), Hourly

Robert D. Caligiuri, Ph.D. 1977 (with Oleg Sherby), Group VP and Principal Engineer

Lawrence E. Eiselstein, Ph.D. 1982 (with Oleg Sherby), Principal Engineer

Eric P. Guyer, Ph.D. 2005 (with Reinhold Dauskardt), Office Director and Principal Engineer

Ricardo Zednik, Ph.D. 2008 (with Paul McIntyre), Managing Engineer

David T. Schoen, Ph.D. 2010 (with Yi Cui), Managing Engineer

Ryan P. Birringer, Ph.D. 2012 (with Reinhold Dauskardt), Managing Engineer

Garrett J. Hayes, Ph.D. 2014 (with Bruce Clemens), Senior Associate

Lucas A. Berla. Ph.D. 2014 (with William Nix), Senior Associate

Chapter 9

Completing the Faculty Expansion

The late 1960s saw the appointment of three more junior faculty members that would bring the faculty head count to eleven, just short of the twelve faculty members that would constitute the Materials Science faculty for the decade of the 1970s and beyond. These appointments included Craig R. Barrett, who had been a student in the department and who was expected to offer courses on x-ray diffraction and spectroscopy and electron microscopy after the departure of Victor Macres; Arthur I. Bienenstock, who was expected to develop teaching and research programs in structures and x-ray scattering; and David M. Barnett, another former student of the department, who was expected to take up the joint position in Materials Science and Applied Mechanics that had been vacated by Barnett's former advisor, Alan S. Tetelman.

Craig R. Barrett ~ From Academic to Captain of Industry

The story of Craig Barrett and his association with the department embodies the evolution of the department in the latter part of the 20th century as well as any story that can be told. From his arrival in the Department of Metallurgical Engineering as a freshman, in 1957, through his undergraduate and graduate studies leading to his 1964 Ph.D. degree in materials science, and his later service as a professor of materials science, Barrett played a key role in helping to establish Stanford's Materials Science Department. After he left Stanford to join Intel Corporation-- where he rose to become the company's president and chief executive officer, and later chairman of the board-- Barrett continued to influence the development of the department with his advice and support.

Perhaps no other graduate of the department has had the kind of impact on technology and society that Barrett has had. As he has explained, although he enjoyed teaching and academic research as a professor at Stanford, he wanted his work to have a greater and more immediate impact. After taking a leave of absence in 1973 to join Intel as a consultant, and joining Intel a year later, he was on his way to helping change the world. With his leadership in the reliable manufacture of memory chips and microprocessors based on silicon technology, he and just a few dozen others were responsible for creating the tools that run the modern world.

Craig Barrett, 1960s. (Stanford News Service)

Craig Radford Barrett was born in San Francisco on August 29, 1939. As a young boy, he moved with his family to the San Francisco Peninsula, first to Hillsdale and then to San Carlos, where he finished his secondary education. He graduated from Carlmont High School in Belmont in 1957. During his senior year, he was reflecting on where to go to college and what he might study. He always had thought he wanted to be a forest ranger, which was a logical extension of his love for fishing and hiking and just being outdoors. He applied to Oregon State, which had a forestry program, and to Stanford, which did not, and was accepted at both schools. He figured it was unlikely he could go to Stanford, as it was exorbitantly expensive, even back in those days, but then Stanford awarded him a scholarship. In Barrett's 2002 oral history interview for the International Archives of the Computerworld Honors Program, Barrett tells a funny story about how he came to major in metallurgical engineering at Stanford. His statement is quoted here:

> I was sitting in my fourth-year math class at Carlmont High School, where I went to high school. My friend Tom Hessler was sitting next to me. I was filling out this application form during class, which I probably shouldn't have been doing, but I was. I got to the point where it said "prospective major," and I knew that Stanford didn't have a forestry department so I needed to have some other major to put down. I elbowed Tom and I said: "What are you going to major in?" Tom was going to go to San Jose State College. He said: "I'm going to major in metallurgical engineering." And I said, "What's that?" He said, "I'm not sure, but I think it has something to do with metals." I said, "How do you spell it? Is that two "L's" or one?" He told me it had two "L's", so I wrote down metallurgical engineering and received a scholarship from International Nickel Company. I was probably the only incoming freshman at Stanford that had ever said that they were going to major in metallurgical engineering. I don't think they had ever given out the scholarship before. So, I was at the right place at the right time. *(Barrett, 2002)*

The Department of Metallurgical Engineering minutes of the faculty meeting of June 3, 1957 do indeed mention that the scholarship provided by the International Nickel Company (INCO) was being awarded to incoming freshman Craig R. Barrett.

At Stanford, Barrett continued the athletic pursuits that he had started in high school, where he played basketball and was on the track team. He joined the track team at Stanford and excelled in several events, notably the broad jump. In a 1961 track meet with the University of California, Barrett jumped 23ft-9in in the board jump, well short of the world's record, of course, but still quite a long distance. (Anyone doubting this should try it.) He also competed in the triple jump and in the low and high hurdles, among other events. In his senior year he received the Irving S. Zeimer award as the most inspirational athlete on the Stanford track team, awarded by legendary Stanford track coach Payton Jordan. Even as a graduate student, Barrett kept up his athletic edge by throwing the discus and practicing the high jump for fun. Nix remembers Barrett regularly clearing six feet in the high jump, still well short of a world's record, but well outside the reach of most casual athletes.

As an undergraduate, Barrett joined Oleg Sherby's research group and helped some of Sherby's graduate students with their experiments. He worked with Jack Lytton and later Alan Ardell on problems involving high temperature creep of metals. After receiving his B.S. degree in 1961, he continued with Sherby as a graduate student, focusing on the effects of grain size and stacking fault energy on the creep of metals and solid solutions. Sherby was building a phenomenological framework for understanding the factors that control the creep of crystalline solids and Barrett's work contributed to that effort. With Alan Ardell, Barrett collected more evidence that the elastic modulus was an important factor to be considered in predicting the creep of metals. In addition, his work on the effect of stacking fault energy led to a new term in Sherby's evolving phenomenological equation.

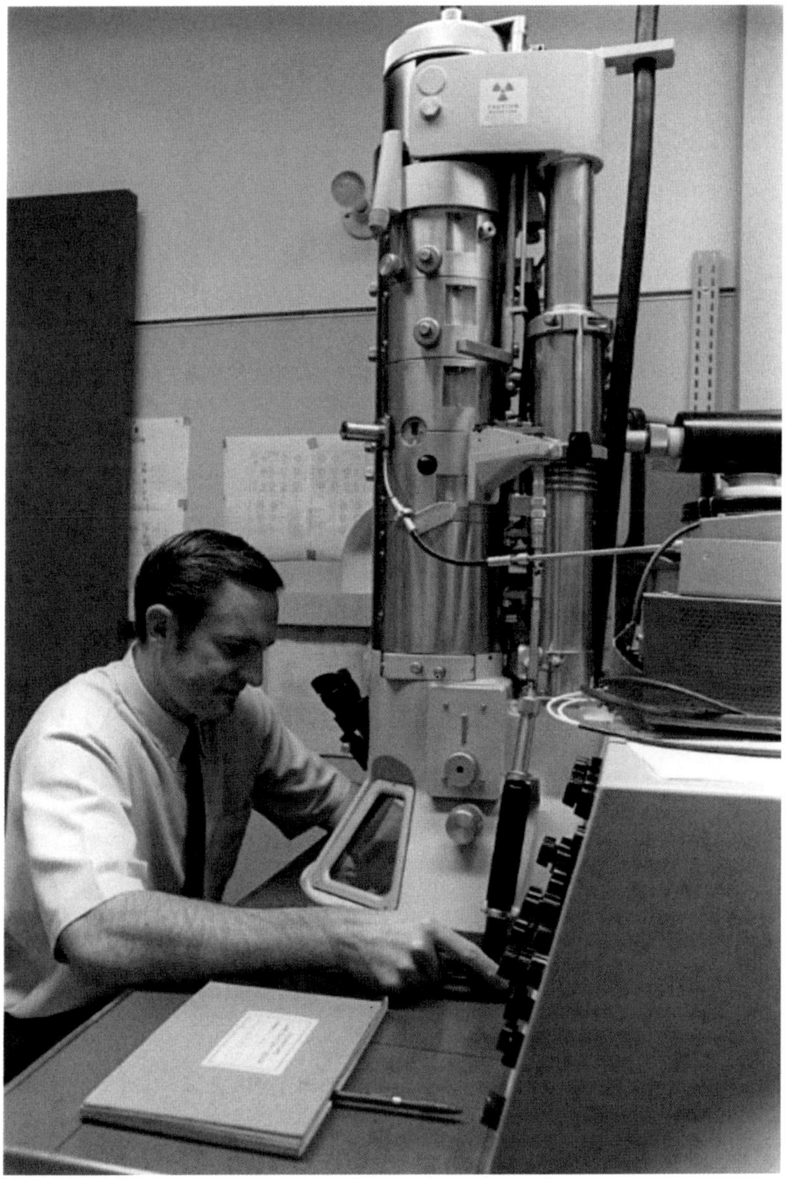

Along the way Barrett also became interested in the theory of high temperature creep and especially in how one might compute the creep rates of metals from knowledge of the diffusional mobility of dislocations and the concentrations of dislocations under different creep conditions. That led him to collaborate with Nix, who was also thinking along those lines. Together they wrote a paper showing how creep might be controlled by the diffusion-controlled drag of jogs by gliding dislocations. This attracted some attention in the field of high temperature creep and became Barrett's most highly cited paper. While the Barrett-Nix theory was ultimately found to be deficient for ordinary metals, it later found

C.R. Barrett posing at the Phillips EM 200 transmission electron microscope in the McCullough Building in the early 1970s. (Stanford News Service)

use in explaining the creep properties of titanium aluminides, which were being developed for gas turbine applications.

After spending a post-doctoral year at the National Physical Laboratory in England under the support of the NATO post-doctoral fellowship, in 1966 Barrett accepted an offer to return to the Materials Science Department at Stanford as an assistant professor. He was brought in to replace Victor Macres, who was then leaving the faculty.

William Nix and Craig Barrett, 1989. (W.D. Nix)

By replacing Macres, who had been teaching in the area of electron optics, it was natural for Barrett to take on responsibilities of teaching x-ray diffraction and spectroscopy and electron microscopy. He had used those characterization techniques in his research and was well prepared for those courses. He also joined in teaching Engineering 50, "Introductory Science of Materials," with Nix and Tetelman, for all three quarters in 1968-69. That collaborative effort led to their writing "The Principles of Engineering Materials," an elementary textbook published by Prentice-Hall in 1973.

Many interesting stories can be told about the writing of this textbook, including much loud talking and shouting in offices and one near fistfight—perhaps not surprising with three strong-minded authors. The whole enterprise was made more difficult when Tetelman left Stanford for ARPA soon after the agreement with the publisher had been signed (Tetelman had to mail in his chapters). He had been the primary driver in finding the publisher for the book and was a key author. (The order of the authors' names was alphabetical but that was a ruse to get Barrett, untenured at the time, to be listed as first author). In spite of the travails in putting the book together, preliminary versions of three different sections of the book were printed in 1968-1970 and these paved the way to final publication in 1973.

By 1967 Barrett had won the Robert Lansing Hardy Gold Medal of AIME, as Huggins and Tetelman had done earlier and as Robert Sinclair would do later.

In research, Barrett continued to work on problems of high temperature creep, partly in collaboration with Nix. He also developed new lines of research on the mechanical properties of brazed joints with his graduate students Harry Saxton and Anton "Bud" West, and made some of the first *in situ* observations showing grain boundary sliding and grain motions during superplastic deformation with Emel Geçkinli. She was the first woman in the department to be awarded a Ph.D. and later had a distinguished teaching career at Istanbul Technical University.

By the early 1970s, Barrett's interests were turning to more applications, and he was consulting with various companies near Stanford where a new industry based on semiconductors was developing. ("Silicon Valley," as a phrase, was not popularly used until the mid-1970s.) After spending a little time with a small start-up company making lead frames for integrated circuit devices, and then taking a sabbatical leave at the Technical University of Denmark in Copenhagen with support of a Fulbright Fellowship, Barrett took a leave of absence from Stanford to test the waters at Intel Corp. Soon thereafter he made the decision to resign from the faculty to pursue his career at Intel as a research engineer.

His work at Intel focused initially on quality assurance in manufacturing. As he has said, his training at Stanford in materials science was a good background for his work at Intel. At Intel, he was surrounded mostly by electrical engineers and chemists who were not especially knowledgeable about the structure and properties of materials. In his first ten years working there, Barrett developed a reputation for creative solutions to difficult manufacturing problems.

In the early 1980s, Intel was on the verge of being swamped by Japanese manufacturers of computer memory chips, and Barrett was asked to find ways to make Intel more competitive with these companies. His approach was to apply statistical quality control and something he called 'copy exactly' in manufacturing. It was an approach he insisted on with evangelistic zeal, often over the resistance of skeptical department managers. The result was a quantum leap forward in the manufacturing efficiency and quality of Intel's computer chips, which helped Intel survive a mid-1980s decline in revenues and led to Intel's revitalization as the best semiconductor company in the world. It also made Barrett one of Intel's stars and was the basis for his election to the National Academy of Engineering in 1994.

Barrett was identified as one of the future leaders of Intel. In 1984, he was made a vice president, and by 1992 he was a member of its board of directors. That led to him being appointed chief operating officer (COO) in 1993 and President and COO in 1997. He was made chief executive officer (CEO) in 1998 and chairman of the board in 2005. On retiring from Intel in 2008, he became chairman emeritus.

The impact that Barrett had at Intel can be shown by considering the growth of the company during his time there. When he joined the company in 1974, Intel was doing about $30M in revenue a year. In 2008, when he retired, the company was doing about $140M a day, a huge rate of growth by any standard. It was quite an accomplishment for someone who started as a freshman in metallurgical engineering at Stanford in 1957, received his undergraduate and graduate degrees in materials science from Stanford, and served on the materials science faculty with distinction for ten years. As he has said: "It was quite a ride."

Since leaving Intel, Barrett has devoted much of his attention to improving education in the U.S. as well as internationally and utilizing technology to raise social and economic standards globally. He serves on (and chairs) many boards and commissions devoted to K-12 education, education in Science, Technology, Engineering and Mathematics (STEM), and improving public awareness of science. When he is not traveling the world for work on these projects, he spends much of his time with his wife, Barbara, at their 26,000-acre Triple Creek Ranch in Montana, where he can be the forest ranger he always wanted to be.

Arthur I. Bienenstock ~ From X-Ray Science to University Service and National Science Policy

Arthur (Artie) Bienenstock joined the Materials Science faculty in 1967 as a tenured associate professor with a joint appointment in Applied Physics. He had started his academic career at Harvard with work on the structure of amorphous materials and was especially interested in the amorphous to crystalline phase transition in chalcogenide glasses, which held potential for a memory technology. He had been encouraged in this work by David Turnbull, who also supported his appointment at Stanford. With his extensive background in x-ray scattering, Bienenstock immediately developed courses in advanced x-ray diffraction as well as courses on the structure and properties of glassy materials, statistical mechanics and phase transitions in solids. Bienenstock was the second faculty member, after G. M. Pound, to be appointed after William Tiller became executive head. Tiller was convinced that Bienenstock's appointment would strengthen the department's offerings in structures.

Arthur Bienenstock, in 1967. (Stanford University Archives)

While the impact of Bienenstock's teaching and research in amorphous materials and x-ray scattering could be seen immediately, his role in building up a world-class Synchrotron Radiation Laboratory at Stanford could not have been foreseen, nor could his leadership in diversifying both Stanford's undergraduate student body and its faculty. And, of course, his national and international leadership in science policy were far in the future at that time.

Soon after he arrived at Stanford, Bienenstock realized that with a tenured position, he could move outside of his own area of research and scholarship to contribute to the workings of the university more broadly. His first effort along that line was as a member of the Academic Council's Committee for Financial Aid (soon after renamed the Committee on Undergraduate Admissions and Financial Aid, with Bienenstock as the first faculty chair). As committee chair, Bienenstock led the campaign to fully eliminate the 1899 limitation on the number of women (to 500) who could register as students at Stanford at one time; although this limit has been lifted by the Board of Trustees in 1933, an unofficial quota (of roughly 1 to 2) appeared to continue in admissions decisions. In 1972, Stanford officially and legally removed any gendered limitations on admissions.

Bienenstock was especially proud to have led that effort, particularly since it was before Title IX. More importantly, it represents his early efforts to provide broad support for women in the university, an effort that he continues to make though his policy advice on fair processes for both the accused and the accuser in university sexual harassment allegations. His work on the Committee for Financial Aid also focused on need-blind admission to the university, and on improving

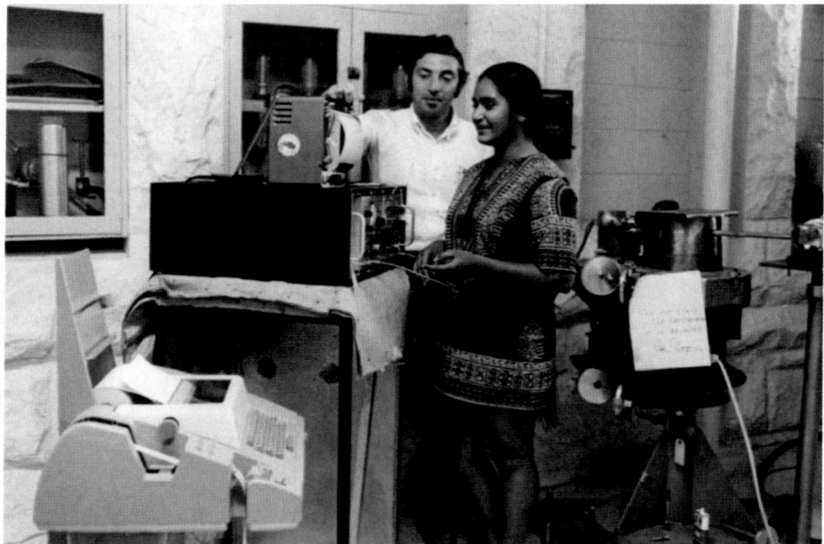

Arthur Bienenstock and Subha Narasimhan working in a lab in the Peterson Building, 1969. (Stanford University Archives)

Stanford's efforts to attract minority students. Bienenstock's work led to his appointment as a University Fellow with a portfolio to look more generally at affirmative action at Stanford and to recommend policies for recruiting highly qualified women and minority faculty. In that capacity, he became the university's Affirmative Action Officer and, subsequently, its vice provost for faculty affairs. In just 5 years from his 1967 appointment, while still in his mid-30s, Bienenstock had joined top administrators at Stanford in bringing about important changes to the university's student body and faculty. He served as vice provost for faculty affairs until 1978, when his new duties as director of the Stanford Synchrotron Radiation Laboratory (SSRL) required more of his attention.

Synchrotron Radiation is electromagnetic radiation in the x-ray, ultraviolet, visible and infrared regimes produced by accelerated electrons when they are forced, by means of bending magnets, to curve from their straight-path acceleration. When the accelerated electrons' path is curved, synchrotron radiation is emitted at the locations where the curving occurs. In the late 1960s, the high-energy physicists at the Stanford Linear Accelerator Center (SLAC) were planning to build SPEAR (the Stanford Positron Electron Asymmetric Ring) for conducting high-energy physics experiments. Since SPEAR was designed to circulate accelerated electrons in an elliptical storage ring, synchrotron radiation was an expected by-product of the storage ring's design, but of no interest or use to SLAC's high-energy physics experimentalists.

William Spicer, who held a joint appointment of both Materials Science and Electrical Engineering, in 1968 proposed to Wolfgang (Pief) Panofsky, director of SLAC, that a port be built at SPEAR to take advantage of the synchrotron radiation for solid-state studies. With partial seed funding from the Center for Materials Research, a station was built at SPEAR in 1972 for materials science/solid state studies using the bright x-ray beams available. That pilot project led to the Stanford Synchrotron Radiation Project (SSRP), a user-based facility with funding from the National Science Foundation and other government and industrial laboratories. From its founding, in 1973, SSRP was initially directed by Sebastian Doniach of the Applied Physics department. Bienenstock became its second director in 1977 when SSRP became the Stanford Synchrotron Radiation Laboratory (SSRL). (In October of 2008, the name of the laboratory was changed again to Stanford Synchrotron Radiation Lightsource.)

It is difficult to overestimate the impact that synchrotron radiation has had on the field of materials science and the enormous impact that SSRL has had on materials research at Stanford. Bienenstock, who served as SSRL director for two decades, from 1977 to 1997, deserves much of the credit for building that facility into a top international resource for synchrotron radiation research.

He is also recognized for his leadership in overcoming the obstacles associated with a laboratory that was, initially, a parasitic operation depending entirely on the needs of an international high-energy physics community. Soon after he became SSRL's director, Bienenstock faced his first major challenge. High-energy physicists were in search of the psi particle and needed to run SPEAR at much lower energies, too low to produce synchrotron radiation in the x-ray regime.

The instrumentation people at SSRL responded to this challenge by developing devices called wigglers: arrays of strong magnets that could be inserted into the straight sections of the storage ring. These would enhance the beam energy enough to produce bright synchrotron radiation where the beam was bent, compensating for the low energies then being used by the high-energy physicists. Still, by running the laboratory as a parasitic operation, the amount of beam time that could be provided to the materials science community was limited to a few months per year, until 1979, when SPEAR became a 50-50 shared facility between SLAC and SSRL. As demand for beam-time at SSRL continued to grow, the decision was made to build a dedicated accelerator to inject high-energy electrons into SPEAR for the explicit purpose of producing synchrotron radiation in the x-ray regime. That development occurred in October 1990, when SSRL took over SPEAR completely, and the high-energy physicists moved their experiments to other, newer and larger storage rings. One of the new, larger storage rings, the Stanford Positron Electron Project (PEP), operated at much higher energies than SPEAR and provided a new source of synchrotron radiation. In 1988 it was fully dedicated to SSRL for one run, when it made use of the newly developed undulators to produce the synchrotron radiation.

Arthur I. Bienenstock, 1986. (Stanford University Archives)

In 1992, SSRL became a separate division of SLAC, completely devoted to synchrotron radiation research. Throughout Bienenstock's 20-year period as director, he acted as both technical and political leader of this effort at Stanford. Without his skillful management of the various travails faced by the laboratory, and without his growing connections to the Department of Energy and the National Science Foundation in Washington D.C., SSRL would have never prospered at Stanford.

Arthur I. Bienentrock was born in The Bronx, New York City, on March 20, 1935. He grew up in the Bronx and attended public schools there, developing an early interest in mathematics and grammar. He skipped several grades along the way and graduated, at age 16, from the Bronx High School of Science. As president of his high school's student organization, he began to develop a talent for speaking in front of an audience, a skill that served him well in his many leadership roles at Stanford and in his dealings with government agencies.

Bienenstock enrolled at the Polytechnic Institute of Brooklyn (Brooklyn Poly) with the aim of becoming an engineer on a kibbutz in Israel, something that had grown out of his interest in the Zionist movement. He soon became more interested in physics, as he became disillusioned about living on a kibbutz. His studies in the Physics Department resulted in his receiving his B.S. degree in 1955, and M.S. degree in 1957. Along the way, he worked in the x-ray diffraction laboratory and, as a first-year graduate student, took a class from Paul Ewald, one of the fathers of x-ray diffraction and chairman of the department. Bienenstock later published two papers with Ewald before leaving Brooklyn Poly to pursue his doctoral degree at Harvard. (Ewald had urged him to leave Brooklyn Poly for a better school.)

Between graduating from Brooklyn Poly and starting a doctoral program at Harvard, Bienenstock spent two summers at the National Bureau of Standards, where he developed his long-standing interest in determining the structure of glass using x-ray diffraction. Although he did not pursue that line of research as a graduate student, he returned to it when he came to Stanford.

Arthur I. Bienenstock, 2013. (Stanford News Service)

At Harvard, Bienenstock worked on the electronic structure of alkali metals and was advised by Professor Harvey Brooks, then dean of the School of Engineering and Applied Physics. Brooks was well known as a leader in both science and science policy, having served on a Science Advisory Committee in the Eisenhower administration. By this time, Bienenstock had started to write papers about his work and to give talks at conferences. On his return from such conferences, Brooks would never ask Bienenstock about how his talk was received. Instead he would always ask: "Who did you meet" and "What did you discuss"? It was as if Brooks was grooming Bienenstock for the career in science policy that he was later to have.

After receiving his Ph.D. in Applied Physics from Harvard in 1962, Bienenstock spent one year as a National Science Foundation Post-Doctoral Fellow at the Atomic Energy Research Establishment in Harwell, England. Following a plan developed before leaving Harvard, he then returned to Harvard as an assistant professor in the Division of Engineering and Applied Physics. His discussions at Harvard with David Turnbull about the structure of amorphous materials rekindled his interest in that subject, and especially in using x-ray diffraction to determine that structure. That work ultimately led to his appointment at Stanford.

While Bienenstock quickly assumed broader leadership roles at Stanford, he also built up a strong group of graduate students, especially in the late 1970s and throughout the 1980s. Much of their work focused on the determination of the arrangements of atoms in polyatomic amorphous alloys by x-ray scattering.

By the mid-1990s, Bienenstock had become well known in Washington D.C., as a result of both his long association with the National Science Foundation and the Department of Energy and their funding of SSRL, and through his role in rounding up support for the National Major Facilities Initiative (synchrotron radiation, neutron scattering, high energy and nuclear physics facilities) that was taking shape at that time. It was not a complete surprise when, in 1997, he was asked to serve as the associate director for science in the White House Office of Science and Technology Policy (OSTP). He served in that capacity until President Clinton left office in 2001.

While working at OSTP, Bienenstock was a strong advocate for the support of the physical sciences, not only for the usual reasons of defense, energy, and transportation, but also for national health. At the time, funding for The National Institutes of Health was to be raised at a far higher rate than for the National Science Foundation and the Department of Energy. Using his background in x-ray scattering, Bienenstock developed a powerful narrative regarding the benefits of x-rays to human health, from the earliest uses of x-rays in medicine to the latest CT scan technologies. He traced the contributions made by mathematics, physics, computer science, materials science, and other fields to this development, successfully making the case for stronger support of the physical sciences, even if national health were considered the top priority for the nation. President Clinton subsequently announced equal raises for all three organizations, with a discussion of the importance of "interdependencies of sciences." Bienenstock played a key role in developing that policy.

On his return to Stanford in 2001, Bienenstock returned to leadership roles, first as director of the Geballe Laboratory for Advanced Materials (2002) and later as vice provost for research and graduate policy (2003-2006). Since 2006, he has held the position of special assistant to the president for federal research policy, and since 2008, he has also served as associate director of the Wallenberg Research Link at Stanford University. In 2012, he was selected by President Barack Obama to serve as a member of the National Science Board of the National Science Foundation. His appointment to the National Science Board was the culmination of a lifetime of service to the nation in science policy and in the science and technology of synchrotron radiation.

Bienenstock has collected an impressive array of honors for his work in science and science policy. He received the Sidhu Award of the Pittsburgh Diffraction Society in 1968, the Distinguished Citation of the Polytechnic Institute of New York Alumni Association in 1977, and the Rector's Lecture and Medal from the University of Helsinki in 1994. He received the Distinguished Contribution to Research Administration Award from the

Society of Research Administrators in 2000 and the Distinguished Service Award of the U.S. Department of Energy in 2005. He was awarded an honorary Ph.D. from the Polytechnic University of New York in 1998 and from Lund University in 2006. In 2009, he was awarded the coveted Kenneth M. Cuthbertson Award for Exceptional Service to Stanford University.

Bienenstock is a fellow of the American Association for the Advancement of Science, the American Academy of Arts and Sciences (2013), the American Physical Society, the California Council on Science and Technology (2009) and the Institute of Physics. In 2010, he was elected a Foreign Member of the Swedish Royal Academy of Engineering Sciences.

In 2018, Bienenstock received an award that recognizes the whole of his work in science. He received the 2018 Philip Hauge Abelson Prize, given by the American Association for the Advancement of Science (AAAS). The award recognizes individuals who have "made signal contributions to the advancement of science in the United States." AAAS recognized Bienenstock for his academic leadership, his efforts to improve public understanding of science and his promotion of diversity and inclusion in science, technology, engineering and mathematics fields.

David M. Barnett ~ A Connection to Applied Mechanics

David M. Barnett joined the Materials Science faculty in the fall of 1969. He had received his Ph.D. from the department two years earlier and had spent one year as a post-doc in Germany with a second year at the Ford Scientific Laboratory in Michigan before returning to Stanford. His appointment was made possible when Alan Tetelman, his doctoral advisor, resigned in 1968 to accept an appointment at UCLA. Barnett was appointed jointly in the Department of Applied Mechanics, which was then a stand-alone academic department. (It later became the Division of Computational and Applied Mechanics in the Department of Mechanical Engineering.) Barnett served as chair of that division in later years.

Barnett had been trained as a mechanical engineer at Rice University and was always attracted to mathematical/theoretical problems. His doctoral dissertation with Tetelman involved work on the elastic interactions of dislocations with inclusions in solids, as a part of Tetelman's overall effort on fracture of composite materials. During his post-doctoral work, Barnett focused on the mechanics of anisotropic elastic inclusions, becoming an authority on the powerful methods of analysis that J.D. Eshelby had developed in the 1950s. This background prepared him well for his first teaching assignments in materials science at Stanford, which included: "Mathematical Methods in Materials Science," "Dislocations in Crystals," and "Fracture of Solids." Barnett's course on "Mathematical Methods in Materials Science" was to become one of the core courses in the graduate program in Materials Science for many decades. In the course of his time on the faculty at Stanford, Barnett made a number of seminal contributions to the development of integral formalisms for solving problems in anisotropic elasticity, which he used to make fundamental breakthroughs in the understanding of defects and interfacial waves in solids. He trained 20 doctoral students in these and related fields. From his appointment to the faculty in 1969 to his retirement in 2015, Barnett served on the MSE faculty for more than 45 years, longer than any other faculty member.

David Barnett was born in Washington, D.C., on November 16, 1939, to Herbert and Selma Barnett. His father, the son of Lithuanian Jewish immigrants, had been born in Chicago; his mother had immigrated from Russia to Austin, Texas, at the age of eleven. The onset of the Great Depression prevented both parents from entering college,

David M. Barnett with another famous math teacher, Jaime Escalante. (David Barnett)

and they later met while both were working in Washington, D.C. In 1941, Barnett and his parents moved back to Austin, where his father opened a men's clothing store on Congress Avenue, a short walk from the state capitol building.

Barnett was educated in Austin's public schools and graduated in 1957 from Stephen F. Austin High School. Having worked in his father's store as a salesman, and for the Austin American-Statesman covering junior high sports, he'd concluded that a career in either retail sales or journalism was not for him. Fortunately, he had been blessed with excellent math teachers, and when Sputnik produced the national need for scientists, engineers, and mathematicians, those fields seemed to be obvious career choices. He applied to the University of Texas and to Rice University (which was tuition-free at that time), was admitted to both, and with the award of a $300 freshman scholarship, opted to attend Rice.

At the end of his sophomore year at Rice, Barnett declared Mechanical Engineering with a metallurgy option as his major, largely due to his interactions with Professor Franz R. Brotzen. Brotzen was instrumental in Barnett's being awarded a $500 American Society for Metals scholarship for his junior year, which helped to bring him to the field of materials. He received his B.A. degree in Mechanical Engineering in a four-year program, staying for an additional two years to receive his M.S. degree in 1963. He continued at Rice for another year, expecting to study there for a Ph.D. in metallurgy, but after attending a graduate seminar on fracture of solids given by Professor J. N. Goodier from Stanford, and learning from Goodier that there was an opening in Stanford's doctoral program for someone interested in both solid mechanics and materials science, he changed his plans. He was admitted to the Department of Materials Science in 1964 and awarded a Ford fellowship to work with Alan Tetelman.

At Stanford, Barnett was shaped as much by his study with mathematicians as he was by his research with Tetelman. He credits Professors Harold Levine, Gordon Latta, and Menahem Max Schiffer of the Stanford math department for training him in the broad field of mathematical physics, including Green's functions and their applications in science and engineering, and variational and operational calculus. His studies of these subjects prepared him well for a lifetime of making elegant contributions to the mathematical theories of mechanics and materials science.

During his graduate studies, Barnett was inspired by a seminar given by John Dundurs of Northwestern University on solutions for edge dislocations near circular inclusions. That gave him the idea of treating cracks and slip bands as continuous distributions of screw dislocations near second phase particles or inclusions. That led to his single-authored paper on the effect of shear modulus on the stress distribution produced by a planar array of screw dislocations near a bi-metallic interface. He was especially proud of that work because it was an unsolved problem that

David Barnett with his wife Muriel and, at right, John Stephens. (David Barnett)

had been attempted by other prominent theorists. It also led to his treatment of screw dislocation arrays near circular inclusions with finite rigidity. The solutions provided in that second paper, and in his Ph.D. dissertation, were included in the recent textbook *Imperfections in Crystalline Solids* by W. Cai and W.D. Nix as the most effective way to treat the interaction of dislocations with inclusions with differing elastic rigidity.

Following his Ph.D. in 1967, Barnett received a NATO Post-Doctoral fellowship to study for one year with Professor Ekkehart Kröner in the Theoretical Physics Institute at the Technical University of Clausthal in West Germany. It was there that he began to thoroughly digest Eshelby's classic 1961 paper on inclusions and inhomogeneities, a process greatly aided by his earlier study of Green's functions at Stanford. A good understanding of Eshelby's work was to be of great utility in Barnett's later teaching and research. Before the end of his post-doctoral studies, Barnett had received several job offers, including one as an assistant professor in engineering from Brown University. He felt he was not yet ready to compete successfully in the academic arena, however, and accepted a staff scientist position at the Ford Scientific Laboratory in Dearborn, Michigan, where the research environment was inviting, and no research fund-raising was required.

The Ford Scientific Laboratory was a most stimulating place, and Barnett had many good chances to discuss questions of mathematics and physics with the many excellent scientists there. There were, however, ominous signs that changes were coming. A short time after Barnett's arrival at Ford, the director of the laboratory was dismissed. In addition, rumors began to surface that the Ford Labs were heading toward increased applied research (much like the trajectory of other industrial research laboratories of the time), and many of the lab's scientists began to make connections with applied groups in engineering. The laboratory management made no effort to suppress these rumors, and conversations in the coffee room were more demoralizing than Barnett wanted to hear. When he learned about the possibility of returning to Stanford to take the faculty position to replace Alan Tetelman, he responded with enthusiasm. He re-joined the department in September of 1969, this time as an assistant professor.

Since Barnett held a joint appointment in Applied Mechanics (later a division of the Mechanical Engineering Department), he taught graduate courses involving elasticity theory, waves in elastic solids, Green's functions for engineers, and variational methods in solid mechanics, in addition to teaching in Materials Science. He also taught a one-quarter introductory course in Statics every year for 20 years, as a part of the undergraduate Engineering Fundamentals series at Stanford.

Although Barnett attracted his share of graduate students in materials science and mechanics and maintained a small group of students interested in theoretical problems, he always felt that he needed to conduct his own theoretical work more or less independently of his student's projects. His view was that the most challenging theoretical problems were not especially suitable for graduate students, as they were both too difficult and not good vehicles for learning to do research. Soon after he joined the faculty, he began to pursue his research on the theory of anisotropic linear elasticity as it applied to dislocations. With the numerical assistance of one of Marsh Pound's graduate students, Lee Swanger, he authored a paper in 1972 that showed how a single real integral (called the Integral Formalism) could be used to determine each component of the tensor elastic Green's functions and their first two derivatives. These quantities form the basis for anisotropic elastic computations involving dislocations and inclusions, as well as other anisotropic elastic problems. His seminal paper laid the foundation for a series of important papers by him and his collaborators, and is the basis for his reputation in mechanics and materials science.

Prior to Barnett's work, solutions to problems in anisotropic elasticity made use of a method that had been developed by A. N. Stroh called the Matrix Formalism. The Matrix Formalism is a powerful and elegant method that relates the static and dynamic properties of anisotropic elastic solids to the terms in a six-dimensional matrix. But a limitation of the method is that the terms in the matrix (which are roots of a six-order

David Barnett, Alexei Maradudin (U.C. Irvine) and Johannes Weertman (Northwestern) at a 1991 symposium celebrating the 60th birthday of Jens Lothe, University of Oslo. (David Barnett)

polynomial) are very difficult to find. This severely limits progress in this area of theoretical research, as most problems involve taking derivatives of these already hard-to-find quantities. With the development of the Integral Formalism, Barnett showed that all the properties of interest in anisotropic elasticity could be written as one-dimensional integrals. He also showed that all the properties of the Matrix Formalism can be re-expressed in terms of the Integral Formalism, and in many cases, doing so leads to not only mathematical simplifications, but also clearer physical insight.

It was Barnett's paper on the Integral Formalism that led to his contact with Jens Lothe at the Institute of Physics at the University of Oslo. Lothe had approached Barnett about the possibility of using the Integral Formalism to study surface waves on anisotropic elastic half spaces. Within days of Lothe's inquiry, Barnett had responded with a treatment of surface waves using his Integral Formalism that soon led to their first joint paper. It was in that 1973 paper where the method was first explained in its full form and discussed in relation to the Matrix Formalism. Barnett and Lothe continued with a follow-up paper concerning the question of existence of Rayleigh (surface) wave solutions, which was still an open issue, i.e., do "forbidden directions for Rayleigh wave propagation exist?" The second Barnett-Lothe paper was finished in January 1974, when Barnett took a six-month sabbatical leave to work with Lothe at the Physics Institute in Oslo. That work was the start of a decades-long collaboration that Barnett had with Lothe on the existence of all types of surface (Rayleigh) and interfacial (Stoneley) waves in anisotropic elastic solids and the conditions in which they exist. This body of work is widely acclaimed as a tour-de-force and a triumph of the Integral Formalism. Their 1974 paper contains what is now called the Barnett-Lothe theorem,

David M. Barnett, 2012. (Stanford News Service)

which is a celebrated result that describes the interaction between dislocations and a bimaterial interface.

The Integral Formalism also allowed Barnett, as well as his collaborators and followers, to make fundamental break-throughs in understanding dislocations and other defects in anisotropic solids, surface and interfacial waves, and piezoelectric materials. In 1975, Barnett and Lothe showed that the six-dimensional Matrix Formalism can be extended to an eight-dimensional formalism to treat anisotropic piezoelectric materials, and great simplification can be obtained by rephrasing everything in the Integral Formalism. This work has attracted considerable attention, eventually leading to his most highly cited paper on that subject with Zhigang Suo and others. His 1978 paper with David Bacon and Ronald Scattergood on defects in elastically anisotropic solids became a standard reference for anyone interested in the elastic fields of defects in crystals.

In addition to his personal scholarship and his work with his own students, Barnett has a reputation for helping other faculty and students work through mathematical and theoretical problems that are difficult for them. Many have said that they could not count the times that they had gone to him for help with what seemed to them like difficult problems, only to find he could construct very neat and elegant solutions, seemingly off the top of his head. His remarkable ability to see through mathematical and physical problems and then come up with elegant solutions is admired by students and colleagues alike. It is this special trait that led him to devise the ingenious solutions for which he is well known.

Barnett's work was recognized by his peers when he was selected to be the 1985 Midwest Mechanics Lecturer. That lectureship involves giving seminars at ten mid-western universities over two one-week periods. In 2012, he was awarded the A. Cemal Eringen medal of the Society of Engineering Science at their annual meeting in Atlanta, Georgia, and treated to a symposium in his honor. And in 2018, Barnett received the Drucker Medal from ASME "for the development of the integral formalism for solving problems in anisotropic elasticity, which he used to make fundamental breakthroughs in the understanding of defects and interfacial waves in solids." Barnett retired from active teaching and research in January 2015 and became an emeritus professor.

In his own 2016 recollection of his career for the author of this piece, Barnett said that writing it "forced me to again recognize that the real value of work and life lies in the people one interacts with along the way, and not with anything one may or may not have accomplished. I got to meet and know many of the very best scientists of my day, and, for a kid who grew up among the hills of Central Texas, that was far more than I could ever have expected. I was blessed with wonderful teachers and good friends at every station and institution along the way." He also said that "Stanford was extremely nice to me, and though I don't think I was overpaid, I would never tell them that." To paraphrase the great Brooklyn Dodger catcher Roy Campanella, "They never knew I would have played the game for nothing!"

Chapter 10

Visiting Professors and Lecturers and the Intimate Nature of the Department of the 1960s

While the department had started to grow rapidly in the 1960s, it was still comparatively small and retained some of the characteristics of a small academic unit. The number of regular faculty members in materials science was still too few to provide all of the teaching and advising needed for a department having both an undergraduate program and a graduate program offering both Master of Science and Doctor of Philosophy degrees.

To make up for the limited number of teachers available, the School of Engineering provided funding that allowed the department to invite visiting professors from around the world to come to Stanford to help in teaching. In addition, the department appointed a number of scientists working in local companies as instructors to offer specialty courses to supplement the basic curriculum offered by regular faculty.

The small size of the department in those years also fostered an intimate academic environment that was family-like in its character. That character was reflected in the day-to-day interactions among faculty members and in the warm social relationships that developed between students and faculty.

Visiting Professors ~ A Rich Resource

The remarkable growth of the department in the 1960s and the rapid development of a national, if not international, reputation has already been mentioned. That transformation was initiated by Shepard and Huggins and furthered by the first faculty members they appointed. It was also made possible by the funding brought to Stanford in the early 1960s and by the new facilities that funding enabled. But the ascendance of the department can also be traced to that time period's visiting professors. A listing of those visiting professors is now a virtual Who's Who in the field of materials science. Among many other prestigious awards garnered over the years, the group's members include three fellows of the Royal Society of London (FRS), two members of the U.S. National Academy of Sciences (NAS), four members of the U.S. National Academy of Engineering (NAE), one member of the Royal Dutch Academy of Sciences, one member of the Polish Academy of Science, and one member of the Indian National Science Academy. None of the visiting professors had yet achieved distinction when they were at Stanford, so credit must be given to the early faculty members for their foresight in bringing such promising people to the department.

The first visiting professor was **Robert G. Ward**, a British metallurgist who was in the department in 1960. His visiting professorship was funded by a Ford Foundation Grant to the School of Engineering to bring distinguished scholars to Stanford. Ward was later the Stelco Chair of Metallurgy in the Department of Metallurgy and Metallurgical Engineering at McMaster University, where he established collaborations with the steel industry that led to the founding of the McMaster Steel Research Center.

Robert Cahn (FRS) was a visiting professor in 1961. As a student of Egon Orowan, he had already made a name for himself by establishing the Cahn-Nye relationship to describe certain aspects of recovery of deformed metals. He became known as the most erudite materials scientist of his time, authoring and editing highly respected books and monographs in materials science. His book entitled *The Coming of Materials Science* is a must-read for anyone wishing to understand how the field of materials science developed in the 20th century.

Ray Smallman (FRS) was a visiting professor in 1962 from the University of Birmingham, where he had been a student of Alan H. Cottrell. He later published books on physical metallurgy and applied electron microscopy and analysis of defects in crystalline materials. He is remembered for loving the game of golf and especially the Stanford Golf Course. In December of 1962 Smallman was offered a permanent faculty appointment, which he declined.

Charles S. Barrett (NAS) from the University of Chicago and **Ted Massalski** from the Mellon Institute in Pittsburgh (Later Carnegie Mellon University) were visiting professors in 1963. They were preparing a third edition of Barrett's classic work, *The Structure of Metals,* first published in 1943. Barrett was already a distinguished professor at that time, having pioneered the use of x-ray diffraction and electron microscopy in metallurgy. He was known for developing the Berg-Barrett x-ray topographical technique for imaging defects in crystals. He was elected to the National Academy of Sciences in 1967. Massalski, born in Poland, was in the midst of a brilliant career in metallurgy and materials science at that time. He was later editor-in-chief of ASM's *Binary Phase Diagrams,* one of the great contributions to the materials field, and later served for more than 40 years as editor of the influential *Progress in Materials Science.* Massalski was later honored with many gold medals and was a member of several foreign academies.

Australian-born **Robert Honeycombe** (FRS) was a 1964 visiting professor from the Department of Metallurgy at the University of Sheffield. Just two years later, in 1966, he accepted the appointment as the Goldsmiths' Professor in Metallurgy and Head of the Department of Metallurgy at the University of Cambridge, a post he held for 18 years, longer than any other head, before or since. Although an authority on steels, he is credited with transforming the Cambridge department from metallurgy to materials science. He was knighted in 1990 for his services to materials science and elected as a Fellow of the Royal Society and of the Royal Academy of Engineering.

Angus Hellawell, who was to become an authority on solidification processing, started his academic career as a visiting professor in the department in 1965. He arrived fresh from his degrees in metallurgy from the University of Oxford. He was later professor of materials science and engineering at Michigan Technological University, where he the received the Bruce Chalmers award for "outstanding contributions to the science and/or technology of materials processing" from TMS.

The year 1966 brought both **Arthur J. McEvily,** from the Ford Scientific Laboratory in Dearborn, and **David Leiberman,** from the

University of Illinois, to the department as visiting professors. McEvily worked with Tetelman on their book, *Fracture of Structural Materials*, which was published in 1967. He and Tetelman had been colleagues at Ford before Tetelman came to Stanford. McEvily went on to have a distinguished academic career in metal fatigue at the University of Connecticut. David Lieberman was best known for his work on shape memory alloys and the physical properties of crystals. He had worked with T.A. Read at Columbia University, where they created the Wechsler-Lieberman-Read theory of martensitic phase transformations, before they both moved to the University of Illinois in 1953. He remained a part of a very strong group of University of Illinois faculty working on martensitic phase transformations.

Morrie Fine (NAE) from Northwestern University, the person who is widely credited as being the father of the academic field of materials science, was a visiting professor for the academic year 1967-1968. Fine and his colleague Donald Whitmore had established the Graduate Department of Materials Science at Northwestern in late 1958. While in the department, Fine wrote a review article extending the principles of precipitation hardening from metals to ceramics. He was elected to the National Academy of Engineering in 1973.

The department also hosted the famous Belgian scientist **Severin Amelinckx** as a visiting professor in 1967. He had already written his authoritative review article, "The Direct Observations of Dislocations," for the *Solid State Physics Series* in 1964. Following his stay at Stanford he continued to make major contributions as director general of the Research Centre for Nuclear Energy in Mol, Belgium. He was later elected to the Royal Academy of Sciences in Belgium and to the Royal Dutch Academy for Sciences, among many other honors.

Albert G. Guy, a professor from the University of Florida who had studied at Carnegie Tech and was best known for his textbooks on physical metallurgy and materials science, was a visiting professor in the department in 1968. He spent his time in the department working on his first book on materials science, *Introduction to Materials Science*, published by McGraw-Hill in 1972. He had started publishing textbooks on materials in the 1950s and continued doing so for the rest of his career, including one with John Hren, one of Sherby's early students.

During the academic year 1968-1969, both **John P. Hirth** (NAE, NAS, AAAS) from Ohio State University and **Jerold Schultz** from the University of Delaware were visiting professors in the department. Hirth had long been associated with Tiller and with G.M. Pound, who had recently arrived from Carnegie Tech, and he would form close research collaborations with Nix, David Barnett, and others, during his one-year stay at Stanford. He was already an authority on dislocations, having published with Jens Lothe, his seminal book, *Theory of Dislocations*, with McGraw-Hill in 1968. He carried a full teaching load at Stanford, teaching courses in thermodynamics, phase equilibria, and statistical thermodynamics. He became one of the most prolific and highly decorated people in the field of materials science and continues his contributions to this day. Several attempts to bring him to Stanford on a permanent basis did not succeed. Schultz also carried a full teaching load at Stanford, teaching courses in crystallography and diffraction. He also taught one of the first courses on polymer science to be given in the department. Schultz spent his entire career in the Department of Chemical Engineering at the University of Delaware.

While the department continued to host visiting professors from time to time in subsequent years, there never again was a focus on strengthening the department's reputation through such appointments. In contrast to the flurry of activity in the 1960s, by 1980 the department was telling the distinguished metal physicist, Frank R.N.

Nabarro, that it could not arrange a visiting professorship due to lack of funding.

In the 1980s, the department did benefit from a visiting professorship program organized by Robert Huggins for the entire materials research community. Huggins, with help from Professor Hans-Joachim Queisser of the Max Planck Institute for Solid State at Stuttgart and the University of Stuttgart, secured a five-year grant from the Volkswagen Foundation to establish the Walter Schottky Visiting Professorship at Stanford. Over the ensuing years the following individuals spent time at Stanford under this program: **Peter Haasen**, University of Göttingen (Haasen, a famous physical metallurgist, was a frequent visitor to the department from the late 1960s to the late 1980s and was once considered for a permanent faculty appointment in the department); **Josef Stuke**, Phillips University in Marburg; **Max J. Schulz**, University of Erlangen; and **Hans-Joachim Queisser**, University of Stuttgart. In 1987, the VW Foundation made an additional grant to Stanford to fund visitors to Stanford for university-industry collaborations.

As an aside, Hans Queisser had worked with William Shockley at his Semiconductor Laboratory in Mountain View, California, starting in 1959. There Queisser worked on building solar cells and also established the Shockley-Queisser limit (which refers to the theoretical efficiency of a solar cell based on a single p-n junction). Queisser maintained long and deep connections to Silicon Valley and Stanford for the rest of his career.

The Lecturers ~ Supplementing the Teaching Program

When the department was being built up in the early 1960s, it was clear that the small number of faculty members could not offer the breadth of courses that would be needed for the department to be competitive with the established programs at other universities. As a consequence, the department adopted the practice of hiring part-time teachers to give courses on materials that could not be covered by the regular faculty.

Claus G. Goetzel was a consulting scientist at Lockheed Missiles and Space Company in Sunnyvale who served as a lecturer in the department from 1962 to 1984. He taught courses entitled "Principles of Powder Metallurgy" and "High Temperature Materials." He had written a five-volume "Treatise on Powder Metallurgy" in the 1950s and had founded Sintercast Corporation of America in New York to exploit his expertise in powder metallurgy. He was also an authority on high temperature materials for Lockheed. Goetzel had been born in Berlin in 1913. He came to the United States in 1936 after receiving his degree in engineering at the Technical University in Berlin and he later received a doctorate at Columbia University in 1939. He served as technical director at American Electro Metal Corporation from 1938 through 1947, where he assisted in the war effort by developing and manufacturing special armor-piercing shells used against the German forces. Following his work at Sintercast, he joined the faculty at New York University's Bronx campus, before moving to Lockheed in 1960.

Donald R. Mash, who had received his Ph.D. from the department in the early 1950s, taught a course on "Nuclear Reactor Materials" from 1962 to 1965. He was working at the Lawrence Livermore Laboratory (later renamed the Lawrence Livermore National Laboratory) where he had developed expertise in nuclear materials. He later became Dean of Engineering at the University of Portland and, still later, a professor at the California Polytechnic School (Cal Poly) in San Luis Obispo.

Leonard Reed, who was manager of the Device Research Laboratory at Eitel-McCullough Inc. in San Carlos, taught courses entitled "Principles of Ceramics" and "Technical Ceramics" in 1963. He was an expert on ceramic-to-metal seals, which were crucial for the construction of vacuum tubes for radios. Eitel-McCullough (Eimac), founded by William Eitel and Jack McCullough, had been among several pioneering West Coast firms manufacturing vacuum tubes before World War II. Jack McCullough was later to provide most of the funding for the construction of the McCullough Building at Stanford, where materials research has been conducted since the mid-1960s.

Donald J. Lyman taught the first course on polymeric materials to be given in the department. From 1965 to 1967, he taught a course entitled "Structure and Properties of High Polymers." At the time he was associated with the Polymer Sciences Department at the Stanford Research Institute in Menlo Park, where his research was focused on biomedical polymers. He later had a long academic career at the University of Utah in both the Department of Materials Science and Engineering and the Department of Biomedical Engineering, where he also specialized in biomedical polymers.

A Picture of the Department in the 1960s

In the course of preparing this history, the author found and read essentially all of the minutes of the faculty meetings from 1957 forward. They provide an illuminating snapshot into the activities of the faculty and the department during that time. As one who was there at the time, and who is still in the department, it is quite surprising to see how much things have changed. One tends to assume that the basic operation of the department has always been about the same. But reading minutes of the faculty meetings shows that that is not at all true.

The most striking thing about the faculty meeting minutes is that faculty meetings were primarily a convenient time simply for administrative information exchange. Before the advent of e-mail or smart phones, the professors were limited to having face-to-face information exchanges at the faculty meetings (when individual telephone calls, informal meetings, and memos to each other weren't inclusive enough).

Faculty meetings were held every week, and every faculty member was expected to attend. And we did. The most junior member of the faculty served as the recording secretary and was responsible for producing the minutes of the meeting. The agenda included many things that are now done behind the scenes by administrative assistants. (Clerical staff at the time consisted mainly of secretaries, who spent most of their time typing memos, course materials, and technical manuscripts.)

Administrative matters, no matter how trivial, were usually handled by professors at the faculty meetings. For example, the progress of individual students in their degree programs is sometimes chronicled in the faculty meeting minutes. These notices included prospective students visits to the department prior to their admission, their admission to the department, their progress through the curriculum, and, finally, their graduation. Various petitions by students regarding their degree requirements were considered by the faculty, and essentially all milestones for individual students, such as passing or failing exams, admission to candidacy, etc., were announced at faculty meetings.

The minutes also are filled with all matters pertaining to courses, their scheduling (rooms and times) and requirements for various degrees. Upcoming visitors, including both prospective graduate students and people from other laboratories and universities, were announced and professors were asked to participate in the visits, and did so. Even small items, such as some piece of equipment needing repair, were discussed and approved by the faculty.

Equally interesting is what was not discussed. The process of hiring of new faculty was apparently not extensively discussed in faculty meetings of the 1960s. There would be announcements that a prospective faculty member might be visiting or that one had been appointed and was arriving, but there was no discussion at all of the parameters of any faculty search that might be conducted or who the candidates were for that appointment. Also, the renovation of the Mining and Metallurgy building and the building of the McCullough building, which both occurred in the 1960s, were not featured in the minutes of the faculty meetings. One has the impression that the operation of the department was more top-down relative to the modern department. (Indeed, Stanford department heads prior to the early 1960s traditionally held the title of executive head, transitioning to the title of chairman (or chair) only in the 1970s.) It would not be until the late 1970s that faculty appointments would be discussed extensively, as recorded in the minutes of the faculty meetings.

The focus on relatively trivial matters at the expense of more important, long-term issues in the 1960s would not be surprising to anyone familiar with typical correspondence of department heads, deans, and even the president a quarter century earlier. In archival records from that earlier period, one finds long letters and memos to and from department heads, deans and the president regarding the academic performance and even the behavior of some individual international students. One also finds the engineering dean writing to emeritus faculty members and others about their use of space.

Another impression one gets by reading the minutes of the faculty meetings is how strictly the various requirements for degrees were enforced. In those days, doctoral students had to pass a comprehensive written examination prior to being allowed to take the oral qualifying examination, which, in turn, had to be passed before an individual was considered a candidate for the Ph.D. An item in the minutes of a 1961 faculty meeting reveals that the written examination in that year consisted of a six-part, six- hour examination taken on a Saturday. An item in the faculty meeting minutes three years later reveals just how tough the written examination could be. Of the 26 students taking the exam in that year, only 15 were said to have passed. An asterisk placed near the names of three who had passed indicated that they "had been warned," presumably about their shaky performance on the written examination. Thus, even written comprehensive exams could be a major obstacle in earning a Ph.D., though not as daunting as the oral qualifying examination, which could strike fear into the heart of any student.

Another item that reflects the times was an announcement by Stanford's new president Kenneth Pitzer about the Christmas holiday schedule for the university in 1969. He announced that Christmas Eve day was to be a full working day for the university, while acknowledging that the day following Christmas day would be a university holiday. One can almost imagine faculty and staff chained to their desks on Christmas Eve that year, a-la Bob Cratchit.

The requirement that doctoral students be proficient in foreign languages had been common in the 1950s, based in part on the fact that much of the technical literature of the time was to be found in foreign journals, especially in German, French and Russian, but also Italian and Spanish. At the end of the decade, the department required a doctoral student to demonstrate proficiency in two foreign languages, usually by appearing before a teacher of that language to show proficiency. In the early 1960s, the requirement was reduced to one foreign language, with proficiency demonstrated by translating a few technical papers of interest from that language to English. With more and more foreign graduate students arriving in the department in the 1960s who already were at least bilingual, however, the requirement for them was set at one foreign language beyond English and their own language. Soon, petitions of all kinds were being submitted, asking for relief from this requirement for various reasons. Finally, in 1971, with still more foreign graduate students in the program, the absurdity of the requirement was admitted and the requirement of another foreign language for foreign students was dropped. Eventually, in 1974, the foreign language requirement for doctoral students was dropped altogether, primarily on the grounds that by that time virtually all journals were publishing papers exclusively in English.

Another department event that gives a sense of those earlier times was the annual department Christmas Party, later called the Holiday Party. The party was a secular event involving music, food, drink and good cheer, as well as a visit by Santa Claus. The small children of some of the people in the department especially enjoyed seeing Santa, who donned the regulation red Santa Claus uniform, complete with a black belt and what looked like black boots, a white beard, and the several pillows needed to mimic Santa's signature girth. A week or so before the party, the students would set up a Santa ballot box near the front office with an unlimited number of write-in ballots. Usually faculty members were the prime targets in the voting, but not always. One year, one of the post-docs was elected, and she was a brilliant Santa.

The voting for Santa was completely corrupt, with the total number of votes cast far exceeding the total number of people in the department. The highlight of the party was the appearance by Santa carrying a big sack of presents that he would give to faculty and staff, after being assured that the person had been good for the year. Usually the presents to the faculty were gags that drew attention to the well-known foibles and characteristics of each person. Santa's presentations were often followed by skits put on by the students in each of the research groups, again often in the form of roasts of their advisors. The skits were very creative and were considered good fun.

One of the earliest and best presentations was by the students of Robert Sinclair, who produced a mock exposé of Sinclair's work on lattice imaging by transmission electron microscopy. Sinclair had shown that the

Santa (played by Bill Nix), with graduate student Wendelin Wright. (W.D. Nix)

periodic arrangement of atoms in a crystalline lattice could be directly observed using TEM. In their skit, Sinclair's students showed a fuzzy image on a TV screen that looked like what Sinclair had been finding. It was two-dimensional array of dots that looked like the regular arrangement of atoms in a crystal. But as the camera zoomed out and the dots appeared smaller and more closely spaced, the observer eventually saw an ordinary ping-pong paddle with the nubs in the rubber coating of the paddle positioned in a regular two-dimensional array, very similar to the arrangement of atoms in a crystal. The implication was, of course, that Sinclair had fabricated his images by photographing the surfaces of ping-pong paddles. The mock exposé elicited a good-natured response by Sinclair and great applause by the assembled crowd.

The Christmas Party tradition was eventually terminated, partly on the grounds that a nominally Christian holiday should not be celebrated by the department (even though the party was always secular) and also on the grounds that students were spending more and more time on their skit presentations and perhaps less and less time in the laboratory doing research. Also, some of the newer members of the faculty were not comfortable with the roasting nature of the skits. In addition, the increasing number of international students, who were not accustomed to satirizing their advisors, seemed to make the event less in touch with the increasingly international nature of the department. The increased research productivity that may have come from cancelling the festive event was judged to outweigh the benefit of fostering amicable and less formal personal relations among students and faculty. This writer is one who regrets the loss of this tradition.

Photoshopped Christmas lampoon of the early 2000s by students of the Dauskardt group used the recently released movie Shrek to rib Bill Nix as the heroic Shrek, with his students: Wendelin Wright as Princess Fiona, Jeff Florando as sidekick Donkey, and Guleid Hussen as the villainous Lord Farquaad. The point of the satire was long ago lost to history. (W.D. Nix)

Chapter 11

Professional Society Connections

By the end of the 1960s, with the department growing in national stature, the faculty played an increasingly prominent role in the organization and operation of the professional societies engaged in materials research. With increased visibility came an increase in the number of prestigious professional society awards to faculty and students that reflected the growing reputation of Stanford's Department of Materials Science.

Founding and Leadership of the Materials Research Society

Leaders of the emerging field of materials research, including Robert A. Huggins, met at The Pennsylvania State University in May of 1973 to establish a new professional society, the Materials Research Society (MRS). By that time, the multidisciplinary field of materials research had transcended the traditional academic disciplines that had once been associated with work on materials. The field's broadened nature could not be fully served by the professional societies that had grown up around certain specific classes of materials, such as metals (American Society for Metals, The Metallurgical Society of AIME), ceramics (American Ceramic Society), polymers (American Chemical Society), semiconductors (American Institute of Electrical Engineers), or composites (Society of Aerospace Material and Process Engineers).

The founders thus envisioned that MRS meetings would be based on multidisciplinary topics in materials research, as opposed to the traditional disciplinary programing common in other societies. Their aim was to address new and exciting problems in materials research regardless of which academic disciplines contributed. The society also was envisioned as a bottom-up society, in which the MRS members themselves determined the direction of the society through their organization of symposia to address the most critical and important issues in materials research. That dynamic structure resulted in many new people with new ideas being drawn into the field. The MRS membership exceeds 14,000 at the time of this writing and includes strong participation not only by academics from universities but also by scientists and engineers from industry and national laboratories.

The following list of founders of MRS reflects the significant role played by industrial scientists, along with academics, from the very beginning:

Eric Baer, Case Western Reserve University

Harry C. Gatos, MIT

Robert A. Huggins, Stanford

Kenneth A. Jackson, Bell Telephone Laboratories

Eric Kay, IBM Corporation

Robert A. Laudise, Bell Telephone Laboratories

Mark B. Meyers, Xerox Corporation

Earl Parker, UC Berkeley

S.V. Radcliffe, Case Western Reserve

Rustum Roy, Pennsylvania State University

B. Sheldon Sprague, Celanese Corporation

Richard S. Stein, Univ of Massachusetts

Sanford S. Sternstein, Renselaer Polytechnic Institute

James J. Tietjen, RCA Laboratories

I. Warshaw, National Science Foundation

Leonard R. Weinberg, Itek Corporation

Students and faculty from the MSE department have been actively involved in MRS from the earliest days. In addition to Robert A. Huggins, who was one of the founders, several MSE faculty and former students have served in leadership positions in the society, including four presidents of the society: John C. Bravman (1994), Merrilea J. Mayo (2003), Shefford P. Baker (2009), and Bruce M. Clemens (2012).

Honors and Awards

Since the early 1960s, MSE faculty and former students have collected an increasing number of honors and awards for their work. When the main focus of the department was on metals, many of the prestigious awards came from either the American Society for Metals (later called ASM International) or the Metallurgical Society of AIME (later called TMS, the Minerals, Metals and Materials Society). Later, as the focus moved away from metals, and as more and more faculty became members of the Materials Research Society (MRS), awards from that society became more common. As many have pointed out, however, the MRS makes many fewer awards to its members compared to the more well-established metallurgical societies, and still fewer when compared to such large and well-established societies as the American Society of Mechanical Engineers. As a result, the MRS awards are even more precious than those from other societies because they are relatively rare. The awards listed below are only those received by multiple people from the department. More awards won by individual faculty members are included in the sections describing their work and career.

The **Robert Lansing Hardy Award** recognizes a young person in the broad fields of metallurgy or materials science for their exceptional promise of a successful career, rather than for any specific accomplishment. The broad fields of metallurgy and materials science include minerals processing, extractive, physical or adaptive metallurgy, and metal processing. The recipeent must be a member of TMS. The award was established by Dr. Arthur C. Hardy in memory of his son, Robert Lansing Hardy, a young man of great promise in the field of physical metallurgy and a Junior Member of AIME, who died suddenly at the age of 25. Faculty and former students who have received the Hardy Medal are listed here, along with their institution at the time they received the award and the year of the award:

Robert A. Huggins, Stanford, 1957

Alan S. Tetelman, Stanford, 1965

Craig R. Barrett, Stanford, 1967

Robert Sinclair, Stanford, 1976

David K. Matlock, Colorado School of Mines, 1975

Edward Goo, USC, 1986

Robert B. Beyers, Stanford, 1987

Glenn Daehn, The Ohio State University, 1992

Shriram Ramanathan, Harvard, 2011

The **Bradley Stoughton Award** for Young Teachers of Metallurgy or Materials Science recognizes and fosters excellence in the teaching of materials science, materials engineering, design, and processing to encourage young teachers in this field. It is awarded by ASM International. Candidates must be 35 years of age or younger. Bradley Stoughton (1873-1959) was President of ASM in 1942. His distinguished career included teaching and administrative positions at MIT, Columbia University, and Lehigh University, where he served as head of the Department of Metallurgy and Dean of Engineering for 35 years. Faculty and former students who have received the Bradley Stoughton Award are listed here, along with their institution at the time they received the award and the year of the award:

William D. Nix, Stanford, 1970

David K. Matlock, Colorado School of Mines, 1979

George M. Pharr, Rice University, 1984

David L. Bourell, University of Texas, 1986

Jeffrey C. Gibeling, UC Davis, 1987

John C. Bravman, Stanford, 1991

Darrell G. Schlom, Cornell University, 1999

Richard P. Vinci, Lehigh University, 2001

Ryan O'Hayre, Colorado School of Mines, 2010

It is common for professional societies to establish the rank of Fellow of that society, an honor that is bestowed on only the most accomplished members. In the fields of metallurgy and materials science, the rank of **Fellow of the TMS the Minerals, Metals and Materials Society** is considered to be the most prestigious Fellow award. It is given for outstanding contributions to the practice of minerals, metals, or materials science and technology and entitles recipients to use the title of "Fellow of The Minerals, Metals & Materials Society." The number of TMS Fellows is restricted to maintain not more than 100 living fellows under the age of 75 at any one time. It is considered a pinnacle award by the society. Faculty and former students who have been elevated to the rank of TMS Fellow are listed here, along with the year of their award:

Robert I. Jaffee, 1972

G. Marshall Pound, 1983

Oleg D. Sherby, 1985

William D. Nix, 1988

John Stringer, 1992

Alan J. Ardell, 1997

Jeffrey Wadsworth, 2000

T.G. Nieh, 2004

Tsu-Wei Chou, 2008

David L. Bourell 2011

Kevin J. Hemker, 2014

Michael J. Mills, 2015

David K. Matlock, 2015

George M. Pharr, 2016

The rank of **Fellow of the Materials Research Society** is similarly prestigious as it recognizes outstanding members whose sustained and distinguished contributions to the advancement of materials research are internationally recognized. The number of new fellows selected each year is capped at 0.2% of the current total professional MRS membership, making this a highly selective honor. The award was established in 2008. Faculty and former students who have been elevated to the rank of MRS Fellow are listed here, along with their institution when they received the award and the year of the award:

Darrell G. Schlom, Cornell University, 2010

Chang-Beom Eom, University of Wisconsin-Madison, 2011

William Nix, Stanford, 2011

George M. Pharr, University of Tennessee, 2012

Cammy R. Abernathy, University of Florida, 2015

Bruce M. Clemens, Stanford, 2015

Yi Cui, Stanford University, 2016

Michael McGehee, Stanford, 2018

The rank of **Fellow of the Electrochemical Society** is another very prestigious honor that has been bestowed on people associated with the MSE department. The award was established in 1989 for advanced individual technological contributions in the field of electrochemical and solid-state science and technology; and active membership and involvement in the affairs of The Electrochemical Society. Since that time the following individuals with connections to the MSE department have been elected as ECS Fellows:

Cammy Abernathy, University of Florida, 2000

M. Stanley Whittingham, Binghamton University, 2004

Eric Wachsman, University of Maryland, 2007

Bryan Chin, Auburn University, 2016

Bor Yann Liaw, Idaho National Laboratory, 2016

Robert A. Huggins, Stanford, 2017

Turgut Gur, Stanford, 2018

Yi Cui, Stanford, 2018

Since the emergence of the Materials Research Society as the dominant society for the academic field of materials science and engineering, the most prestigious award for young scholars has been the **MRS Outstanding Young Investigator Award**. The award recognizes outstanding interdisciplinary scientific work in materials research by a young scientist or engineer. The award recipient must also show exceptional promise as a developing leader in the materials area. At the time of this writing, only three individuals from the MSE department, all faculty members, have been selected for this special honor. They are listed here, along with their citation:

Michael D. McGehee, Stanford, 2007: *"For innovation and application of organic semiconductors in lasers, light-emitting diodes, transistors and solar cells."*

Jennifer A. Dionne, Stanford, 2017: *"For innovating new materials and methods to visualize and control nanometer-scale optical, electronic, and chemical processes in situ"*

William Chueh, Stanford, 2018: *"For groundbreaking research on ionic and electronic charge transport and interface chemistry relevant to electrochemical devices"*

Election to the United States **National Academy of Engineering (NAE)** is perhaps the highest honor an American engineer can receive. At the time of this writing, the following faculty members and former students have been elected as Members of NAE. The list indicates their institution when elected to NAE and the year of the recognition:

Robert I. Jaffee, Battelle Memorial Institute, 1969

Oleg D. Sherby, Stanford, 1979

William D. Nix, Stanford, 1987

Craig R. Barrett, Intel Corporation, 1994

Gerald B. Stringfellow, University of Utah, 2001

David K. Matlock, Colo. School of Mines, 2003

George M. Pharr, University of Tennessee, 2014

Warren C. Oliver, Nanomechanics Inc., 2016

Darrell G. Schlom, Cornell University, 2017

Chapter 12

Women in Materials Science

For most of its history, the department has been, even more than other engineering departments, a male-dominated enterprise. As the centennial of the department approaches, however, women are playing an increasingly significant role in it and in the entire field of materials science. Once a rarity, women now receive around four of the department's Ph.D.'s each year, and two women serve as tenured faculty members. While this is still hardly gender parity (about 18 Ph.D.'s per year are awarded to male graduate students, working with 18 faculty), the journey and contributions of women in the Department of Materials Science is worthy of special mention.

Pioneering Women

From the founding of the department in 1919 to the late 1940s, the degree of Engineer was the most advanced graduate degree involving research. Of the 12 people who received the degree of Engineer during that 30-year period, only one was a woman: Janet Zaph Briggs. Briggs received the degree of Engineer in Mining Engineering in 1933 and went on to a distinguished career as a steel metallurgist at Crucible Steel Company of America. She was later elected to the National Mining Hall of Fame. No other metallurgical degree (M.S. or B.S.) was granted to a woman during that period.

By the early 1950s, when the M.S. and Ph.D. degrees became the main graduate degrees granted by the department, none of the candidates for those degrees were women. It would not be until 1964 that an M.S. degree was

Janet Zaph Briggs (National Mining Hall of Fame)

Alice Fischer-Colbrie, working as a chemist at NASA's Ames Research Center before starting at Stanford in 1974. (Alice Fischer-Colbrie)

granted to a woman, Shahla Radjy, an Iranian woman who worked with William Nix. Sadly, Radjy died of breast cancer just a few months after she received her degree. Mary Ellen Gulden, who also worked with William Nix, was the second woman to receive the M.S. degree, in 1965, but her career, too, was cut short by her early death. But the tide was changing as, gradually, more women took up graduate work in materials science.

The women who received the M.S. degree in that time period were:

Shahla Radjy, 1964

Mary Ellen Gulden, 1965

Ayşe Emel Geçkinli (Fedai), 1969

Marcia Almassy (Yamada), 1970

Charlotte Meredith, 1971

Helene F. Arnould (Netillard), 1971

Subha Narasimhan, 1972

Kirsten Banke, 1972

The undergraduate women who studied materials science at Stanford in the 1960s and 1970s also were pioneers. They survived through their grit and determination in a nearly all-male and all-graduate level department.

The story of Caroline J. Fouts illustrates just how alone the women were in the department at that time. For the 10-year period, from 1967 to 1977, the department granted only four bachelor's degrees, two of them in 1977, one of which was to Fouts.

Fouts had almost no support from undergraduate classmates, let alone other young women on campus, during her time in the department. Nevertheless she continued, receiving her M.S. degree in 1978, and went on to a successful career at Hewlett-Packard There she worked on the strength and fatigue of chemical vapor deposited silica for fiber optic devices before developing an expertise in high tech laboratory design, which she later turned into a consulting business.

Alice Fischer-Colbrie is another student from that time period who showed her own kind of grit and determination as a student. As an undergraduate student in the early 1970s, Fischer-Colbrie supported herself by working nights and weekends as a crystal grower at Siltec, a company that specialized in growing single crystal silicon boules for the semi-conductor industry (Siltec had been founded by Robert Lorenzini, who had received his B.S. and M.S. degrees from the department in 1959 and 1960, respectively). After receiving her B.S. degree in 1979, Fischer-Colbrie went on to receive both her M.S. and Ph.D. degrees from the department, and had a productive career involving applications of synchrotron radiation techniques to materials.

The year 1973 is a special year in this story because that is the year that the first Ph.D.s were granted to women by the department. In spring quarter of that year, Ph.D.s were granted to Ayşe Emel Geçkinli (Fedai) and Diane Roberston (Margel), both of whom had studied with Craig Barrett. Geckinli was a Turkish student who wrote her dissertation on the direct observation of superplastic flow in a scanning electron microscope. She submitted it a little earlier than Robertson submitted hers on the

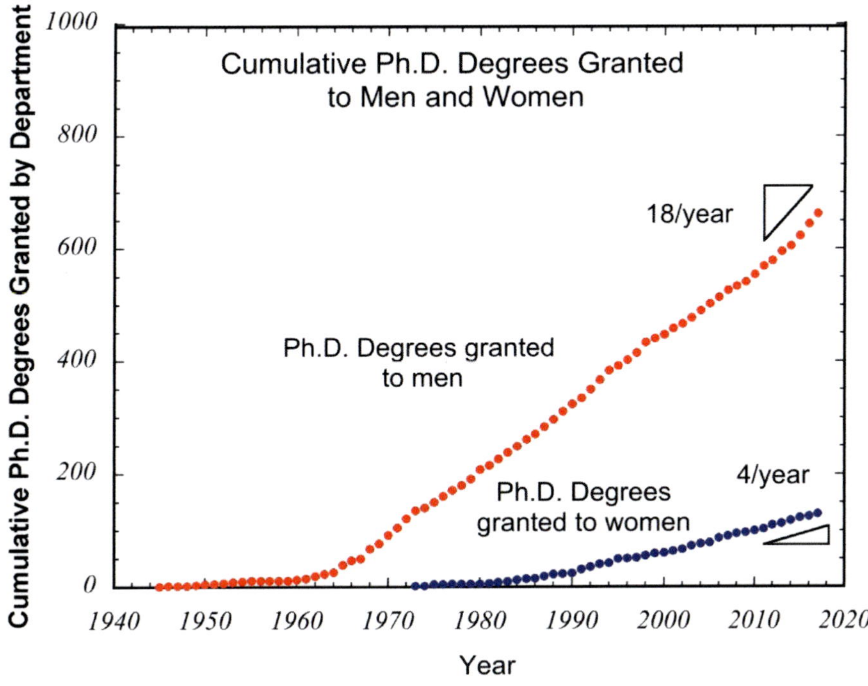

Although these pioneering women had made inroads, the rate of change was still very low. By 1976, four more women had been granted Ph.D.s, but no others received degrees for the following four years. Not until t did the mid 1980s did the number of women receiving Ph.D.s begin to rise significantly. The graph at left shows the cumulative number of Ph.D.s granted to women compared to the cumulative number of men as a function of time. The data shows that for the latest years, the number of men receiving Ph.D.s from the department was about 18 per year while the number of women receiving Ph.D.s was a little more than 4 per year, or about 20%. Parity has hardly been reached but it is within sight.

mechanical properties of bone, so she is regarded as the first woman to receive a Ph.D. in materials science at Stanford.

Geçkinli also gave birth to two children while she was completing her Ph.D., which might qualify as a record itself. She became a distinguished professor at Istanbul Technical University before retiring and continued to teach at another university before her final retirement.

Ayşe Emel Geçkinli, in June 1973 (top) and in 2009 (right). (Emel Geçkinli)

Equally important, the Materials Science faculty did not include women until 2004, when Sarah Heilshorn joined the faculty as an assistant professor. Heilshorn had received her Ph.D. in chemical engineering from Caltech in that year. She was joined six years later by Jennifer Dionne, who came to the department as an assistant professor, after receiving her degree in Applied Physics in 2009, also from Caltech. Both had been promoted to the tenured rank of Associate Professor by 2017.

Prior to Heilshorn's arrival in 2004, the Department of Materials Science and Engineering was the only academic department at Stanford University without any women on its faculty. Needless to say, that was a distinction the department was happy to relinquish in 2004.

The Great Women's Bathroom Caper

Emerging from a male-only department in the 1970s and 1980s brought some challenges that in hindsight might now be seen as humorous. One might call what follows the great "Women's Bathroom Caper." A notice of this (but not by that name) appears in the minutes of the faculty meeting of October 19, 1983, where Nix, as the Associate Chair, was asked to look into the matter and resolve the problem.

First, some background information about bathrooms on the older Stanford campus. In the early days of the university, bathrooms were not located inside the academic buildings. Instead, they were provided in small buildings scattered around the campus, with separate bathrooms for men and women. Some of these bathrooms are still in use. In 1908, at the time the Metallurgy Laboratory was created by adding a wing to the civil engineering building to make an L shaped building, there were no bathrooms included in the building. Students, faculty, and staff were required to use the facilities found in the separate bathroom buildings in the nearby Quad.

When a second new wing was added to the building, in 1915, making a U-shaped building called the Mining and Metallurgy Building, a men's bathroom was included in the new wing. It was still in use in the late 1950s and early 1960s. (As an aside, it included toilets in two separate stalls and one urinal. One of the stalls was reserved for faculty, who used a key to access that stall. Students and others of lower rank were required to use the other stall.) No women's bathroom was available until the L-shaped part of the Mining and Metallurgy building was renovated to form the Peterson Laboratory, in 1963. At this time, several bathrooms were included in the renovated building, including one extremely tiny one for women.

Sarah Heilshorn, 2011.
(MSE Dept.)

The tiny women's bathroom was just down the hall from the office of the department's chairman, near the offices of the administrative and clerical staff. The bathroom's overall size was like that of a broom closet, yet it was crazily partitioned by a wall into two even tinier spaces. This design had an entry space from a much earlier era, like a 1950s "cloak room" with a mirror and coat and hat storage space. Unfortunately, the entry space was squeezed to nothing when the door from the hallway was opened. The second space contained just enough room for two stalls and a sink. The bathroom was so incredibly tiny that if one person was washing her hands in the one sink, another person could not enter the bathroom because the opening door would hit the person at the sink.

The bathroom was also near a big lecture hall in room 550A, which was a favorite venue for big classes from across the university—including classes for majors with many more women than were typical of materials science classes. Before and after these classes, long queues of women could be seen at this, the one and only small women's bathroom in the building. One could see, and hear, a 50 to100-foot-long line noisily decorating the hallway and leading out of the building to the sidewalk. This, plus the growing number of women students in materials science and engineering, and the increasing number of women working as administrators and secretaries, seemed to clearly reveal the inadequacy of that one small women's bathroom. The situation was becoming especially critical in the early 1980s, when the number of women students in the department started to increase significantly.

The problem was brought to a head, in 1983, when Darcy Hughes, Debbie Yaney and Martha Mecartney, with encouragement from Merrilea Mayo, pulled off the "Women's Bathroom Caper." Here I quote from a report on that incident provided to me by Darcy Hughes:

> Inspired by actions of far more noble social causes, we impulsively hatched a plan to take over one of the three men's bathrooms. Three of us women (Darcy, Debbie, Martha) chatted one Friday evening after the department seminar and during the traditional beer gathering in the backroom. As time grew late a heated feminist discussion arose that included our long wait in the post seminar bathroom queue. The last male student with us, Martha's boyfriend, quickly decided to head out, leaving the building just to us.
>
> First, we needed a sign for our takeover and like today turned to a computer with a printer. The technology was so primitive that we had to slowly code each large type face letter: **WOMEN** to print a simple sign. After conducting tours of the men's bathrooms, our takeover choice was the one upstairs which was the smallest and most likely to be handed over. Unfortunately, it was also Professor Sherby's favorite. We placed the paper sign, initiated the takeover, went home and waited for events on Monday. Monday morning: sign gone, no word spoken. Monday evening: a new sign printed on the HP printer and replaced. This exercise in futility went on over and over again until the following Thursday night when extreme measures were taken.
>
> That evening I purchased from my local privately-owned hardware store individual three-dimensional metal letters that closely matched the typeface, size and alloy on every door in Petersen. Bringing small nails, hammer and screwdriver on Friday morning at

Debbie Yaney, William Nix, and Darcy Hughes, figures in the Women's bathroom caper story, 1986. (Darcy Hughes)

> 5:00A.M., I very professionally replaced MEN with **WOMEN**. Note all letters were replaced which gave an official appearance and was symbolic.
>
> Friday morning at 10:00 A.M. Professor Nix appeared in the doorway of our graduate student offices and said: "We have a problem. Professor Sherby has sent me a letter and he is," and here I paraphrase in rude modern language "really pissed." (Hughes, 2017)

As Darcy has reported, Sherby went ballistic over the impertinence he thought had been expressed by the sign and he spent much of the week ranting about the affront. That led to the item in the minutes of the faculty meeting and to the author being asked to resolve the situation. The author found himself in the uncomfortable position of writing a memo about toilets to Stanford's Vice Provost for facilities, describing the situation, and asking for relief. The memo included the facts that the three men's bathrooms in the building included a total of six toilets and six urinals for a total of 12 facilities for men. By comparison, the small women's bathroom included only two toilets. There was a 12 to two disparity of facilities for men and women in the building, a situation that demanded attention.

Fortunately, it was not necessary to explain the biological realities that led to the 12 to two count. Neither was it necessary to explain that the "cycle time" for women is longer than for men. A temporary solution was provided by converting the old men's bathroom in the back of the building to a women's bathroom, by simply hanging a sign on the door. But it was not until the back wing of the building was renovated in 1990, under ADA rules, that a new unisex bathroom was provided.

Even then, the department was nowhere close to "potty parity," given the numbers and realities just described. Sad to say, "potty parity" is almost never reached even in modern buildings designed by progressive, open-minded groups, as indicated by the disparity in the lengths of the queues that are seen outside these facilities today. But we can hope that the great "Women's Bathroom Caper" of 1983 moved us slightly closer to achieving departmental "potty parity."

We can report, however, that the women in this story, like many of the more than 150 women who have earned Ph.D.s in materials science at Stanford by the time of this writing, went on to have very distinguished careers in materials science and engineering and, hopefully, were able to work in more comfortable building accommodations. Their scientific and engineering contributions made in academic, industrial and government institutions helped create the vibrant field of materials science and engineering we find today. In that sense they followed in the footsteps of Janet Zaph Briggs, who started it all so many decades ago and set the path for so many more women to follow.

Student at the X-ray photoelectron spectroscope, 1984, from a CMR Brochure. (MSE Dept.)

Chapter 13

Student—Faculty Sports Activities

Department students and faculty, like the rest of the Stanford community, have been engaging in athletic pursuits throughout the history of the department. But prior to about 1960, the department was simply too small to participate in team sports. Typical of that earlier time were tennis matches, golf matches, running, hiking and other ways to enjoy Stanford's extraordinary outdoor environment. But with the rapid growth of the department in the 1960s, team sports became more common. Indeed, the 1960s through the 1980s may have been the department's Golden Age of Sports. Not only was the department big enough to support sports teams, but the slower pace of life in that time period, compared to today, allowed more time for sports activities. There seemed to be less pressure on students and faculty in that earlier time.

Student-Faculty Athletics

In the 1960s and 1970s, it was not uncommon for students and faculty to gather at noon two or three times a week in good weather to play volleyball—which, in Stanford's case, is throughout much of the year. Oleg Sherby was usually the ringleader. He would start rounding up players just before noon. In the summer, the volleyball games would sometimes be followed by a trip to the Palo Alto Creamery, on the

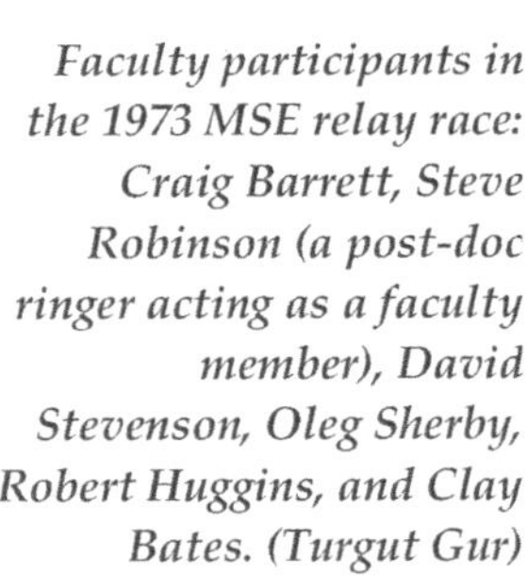

Faculty participants in the 1973 MSE relay race: Craig Barrett, Steve Robinson (a post-doc ringer acting as a faculty member), David Stevenson, Oleg Sherby, Robert Huggins, and Clay Bates. (Turgut Gur)

MSE faculty-student relay race of October 12, 1973. Top row, faculty (left to right): Craig Barrett, Steve Robinson, David Stevenson, Oleg Sherby, Robert Huggins, and Clay Bates. Bottom row, students (left to right): Hugh VanderPlas, Mike Murzik, Jerry Lawrence, Dick Lund, Mike Thomas and Kim Mitchell. (Turgut Gur)

corner of Emerson and High Streets, for lunch. There was also a ping-pong table in the back of the Peterson Laboratory that was heavily used for about an hour after work. Oleg Sherby, William Nix, and David Barnett, as well as many students and post-docs, and even some staff members, would play. Such was life, and "de-stressing," in the department in those earlier times.

While there was never a track team in the department, students and faculty did engage in track and field competitions from time to time. Sometime in the 1960s, a number of graduate students were talking about how they would do in the mile run. After the boasting got out of hand, they found themselves in Stanford Stadium lined up for a mile run without any preparation or warm-up at all. They all finished the race with unimpressive times of 6.5 to 7.5 minutes, but they were completely exhausted and each vowed to limit his future boasting about track and field.

A more serious competition occurred, in 1973, when a group of graduate students challenged the faculty to a relay race in Stanford Stadium (which included, at that time, a full track around the football field). The plan was for six students to compete with six faculty

Handing off the batons in the 1973 MSE faculty-student relay race. On the right, Robert Huggins is receiving the baton from David Stevenson (hidden). Turgut Gur is the timer on the left. In the middle, Mike Thomas has taken the baton from Gerry Lawrence. Oleg Sherby is seen on the far left with his hands on his knees. Although the faculty was ahead at this point, they lost the race. (Turgut Gur)

David Stevenson, National Masters Track and Field, San Diego, 1989. (Family of David Stevenson)

members in a 1.5-mile relay race, with each participant running one time around the track, a quarter mile. Steve Robinson, a post-doc with Sherby, was drafted to join the faculty squad, which was short-handed. Although Craig Barrett, a top athlete by any standard, anchored the race for the faculty, the race was won by the students. The photographs shown here document that event. The student participants included: Hugh VanderPlas, Mike Mruzic, Gerry Lawrence, Dick Lund, Mike Thomas, and Kim Mitchell. The faculty participants were Craig Barrett, Steve Robinson (the ringer), David Stevenson, Oleg Sherby, Robert Huggins, and Clay Bates.

David Stevenson was perhaps the best all-round athlete the department has ever had. He excelled in backpacking, hiking, rock climbing, surfing, skiing, running, biking, and swimming. In the early 1970s, he ran in six marathons and, in the 1980s, began to focus on the triathlon. From 1986 through 1990 he participated in the five Ironman Triathlon World Championships in Hawaii. He won three of the competitions in his age group and was always among the top five finishers in his age group. He recorded his best time at age 61!

By the 1980s, golf had returned as a popular pastime with some faculty and students. In the late 1970s, golf matches included William Nix, graduate students George Pharr, Paul Gilman, and Rick White, and Jeff Wadsworth, a post-doc with Sherby. It was not uncommon for the golfers to joke about hitting the big ball (the earth) instead of the small ball (the golf ball). It was also not uncommon to ask a player who was about the tee off if he inhaled or exhaled on

Michael Mills, William Nix, and John Stephens at Stanford Golf Course, about 1982. (Michael Mills)

MSE entry the Bay to Breakers Race, May 16, 1982. Back row, left to right: Jim Earthman, Michael Mills, Warren Moberly, Jim Proft. Next row: Jose Benzunce (hidden), Subahsh Shinde, Blair London. Next row: Paul or Glen Carry (they were twins and hard to tell apart, especially when partially hidden as in this photograph), John Stephens. Front row: Toshi Oyama and Carry's girlfriend. (Toshi Oyama)

his back swing. The question was intended to distract the golfer by giving him one more thing to think about.

In the early 1980s, the golfers were William Nix and graduate students Mike Mills, Jim Earthman, John Stephens, and Greg Henshall. A favorite story from that period is about the time that John Stephens took a mighty swing at his ball on the 5th tee at Stanford Golf Course, only to see his bladed ball scream off the tee box with an elevation of about 6 inches. About 50 yards out in the fairway was a flock of unsuspecting blackbirds looking for things to eat on the grass. Stephens' ball struck one of the poor creatures, killing him instantly. With his characteristic dry wit, Stephens simply said: "The damn bird killed my roll."

Another golf story from later in that decade was told by Kevin Hemker, who was playing for the first time with his fellow student, Tim Weihs, and their advisor, William Nix. A crusty old starter was guarding a part of the course when George Schultz, then Secretary of State, was trying to tee off. The old starter came running out screaming that that part of the course was closed today. A secret service agent tried to explain that this was Secretary of State, George Shultz, to which the old starter said, "I don't care whose secretary he is, the golf course is closed, and he is not playing today."

The annual Bay to Breakers race in San Francisco was another venue where athletes from the department could be found in the early 1980s. An especially memorable entry was the group that entered in 1982 as a bowling ball with ten pins.

Department Sports Teams

Arguably the best sports team to come out of the department was the late 1960s touch football team called the "Dislocations." The department team competed with fraternities and other living groups in the intramural league at Stanford. The team included Craig Barrett, who was an assistant professor at the time, many materials science graduate students and some graduate students from other departments. In 1968, the "Dislocations" defeated the Phi Delts 12-6 in an overtime thriller to win the intramural touch football championship. The game was tied 6-6 at the end of regulation play and went into overtime. The teams were each given four downs in overtime, with the winner to be determined by the number of yards gained by each in those four downs. On the last play of the Phi Delts series, Craig Barrett intercepted a pass and raced 80 yards for a touchdown to win the game. The "Dislocations" had also outgained the Phi Delts in their four overtime downs. The photograph below shows the 1968 team.

With the increasing number of international students in the 1960s and 1970s, it was not long before the department fielded a team for soccer. A soccer team made of mainly graduate students from the materials science department played in the intramural league in 1969 and 1970. The team included international students Rogelio Gasca-Neri (Mexico), Ricardo Menezes (Brazil) and Tom Surek (Canada) from the materials science department as well as a few students from other departments. As an aside, the team included Jacques Depelchin, who would receive his Ph.D. in African History from Stanford in 1974 and who would go on to become a world-famous Congolese historian and militant.

The department also had a softball team for a couple of years in the early 1970s that played in an intermural league at Stanford. One of the participants from that era reports that while everyone enjoyed the competition in intramural

The "Dislocations," the 1968 intramural touch football champions at Stanford. Participants included: top row, left to right: #4 Craig Barrett, #7 Fred Fehrer and #8 Mike Evans (the quarterback); bottom row, left to right: #1 Tom Read, #2 Bob Maddix (Chemical Engineering), #3 Don Nason, #4 Bud West and #5 Bob Johnson. (W. Robert (Bob) Johnson)

Materials Science soccer team, 1970. Top row, left to right: Tom Read, Leo Amedei, Rob Whalen, Norm Ahlquist, Grant Elliot, Paul Maxwell, and Glen Edwards. Bottom row, left to right: Jacques Depelchin, Bob Asaro, Rogelio Gasca-Neri, Tom Surek, and Jeff Goll. (Amedei, Depelchin and Goll were drawn from other departments). (Grant Elliot)

sports, some of the teams might have done better had not the players argued so much about who would be playing which positions and where they should be in the batting order. Such are the limitations of sports teams made of ambitious young engineers and scientists.

Another intramural softball team was organized in 1976 that included Chuck Schmidt, Richard L. White, Robert Caligiuri, George Pharr, Gary Michal, Paul Gilman, Jeff Wadsworth, and Paul Fuoss. One of the stories from that time involved Warren Oliver's failed attempt to successfully reach first base after he had grounded to third. On his way to first base, he looked over his shoulder for the throw that was coming, only to catch a cleat on the base-path and lose his balance (he was wearing cleats for the first time). Rather than gaining his balance by missing first base, as he should have done, he awkwardly stepped on the base and tumbled head over heels down the first base line, compressing his shoulder and breaking his collarbone in the process. He drove himself to the hospital in his VW van and the game went on without him.

Other softball teams were organized in the early 1980s and we have photographs to document their doings. As can be seen in the photos, many of the same people appeared year after year, leading one to wonder about the progress they were making in their studies and research.

The department fielded a touch football team in 1980. Players from the department included Jim Earthman, Sean Brennan, Mike Kassner, Dan Stern, Toshi Oyama, and Dave Ricky.

Materials Science softball team, 1972. Top row, left to right: Tim Wright (MSE), Paul Pinsky (Operations Research), Leon Johnson (Education), unknown (MSE), Paul Gilman (MSE) and Tim Reiley (MSE). Bottom row, left to right: Eric Hellstrom (MSE), Ron Hanson (Mechanical Engineering), Pat Pizzo (MSE), and John Holbrook (MSE). (Patrick Pizzo)

A department bowling team played in the 1975 fall intramural league. The team included David Barnett, by then an associate professor, and graduate students Paul Gilman, Jay Spingarn, and Jeff Gibeling. Gilman said that Gibeling was a terrible bowler, but he fortunately had an out-of-body experience in the championship game and pushed the team to a win. He said that Ralph Cramden (bus-driving and league bowling TV character played by Jackie Gleason on then-popular TV show "The Honeymooners) would have been proud. Barnett remembers that the team carried quite a high handicap, which probably explains a lot. Gilman did have one game in the season when he had eight strikes in a row, so naturally he has the best memory of the season.

MSE softball team, 1980. Top row, left to right: Steve Galofalini, Jim Smith, Ken Fuoss, Warren Oliver, Dave Rickey, Paul Fouss and Mike Kassner. Bottom row, left to right: Toshi Oyama, and William Nix. (Toshi Oyama)

By the late-1980s, soccer was again becoming popular in the department. Robert Sinclair reported that he and several graduate students had a fairly serious intramural soccer team that reached the finals of its division one year. The team included Sinclair and graduate students Heywood Robinson, Belgacem Haba, Kevin Hemker, Guleid Hussen, Drew Hoffman, Warren Moberly, and Tim Weihs, among others. One of the stories from that season involved

MSE softball team, 1981. Top row, left to right: John Stephens, Bill Avery, Unknown, Kyung-Soo Ahn, Oscar Ruano and Warren Oliver. Bottom row, left to right: Jim Earthman, David Barnett, Toshi Oyama, and Mike Mills. (Seung-Min Han)

MSE touch football team, 1980. Top row, left to right: Jim Earthman, Sean Brennan, Mike Kassner, and Dan Stern. Bottom row, left to right: Toshi Oyama, a woman called Debra, Becky Crites, and Dave Rickey. (Toshi Oyama)

an IM soccer game in the rain. Rain was so rare at Stanford that most of the students showed up wondering if we they were going to play. But when the British professor (Sinclair) showed up, he shouted, "What a wonderful day to play football," and did a belly flop/slide in a big mud puddle. The students followed him in and had a great time. No one can remember if the game was won or lost, but it was memorable.

By the 1990s, the era of MSE participatio n in intramural team sports had largely passed, with students and faculty pursuing athletic activities on their own, much as was the case in the early days of the department. As a postscript on this piece, the sporting activities of John J. Stephens, who is mentioned in the stories above, should be highlighted. Stephens, who had great enthusiasm for all kinds of sporting activities, had been afflicted with Cystic Fibrosis since birth, though most did not know about his illness. He had a reputation for being a very warm and helpful colleague who was always ready to lend someone else a hand. Indeed, he was an inspiration to his colleagues. When he finally succumbed to the disease at the age of 51, most of those with whom he had worked and played were shocked to discover that he had been ill at all. A graduate student award was created in his honor and is given each year to recognize "an outstanding student who had inspired others most." Later, Stephens' mother, just before she died, endowed the John Stephens, Jr., Graduate Fellowship Fund in memory of her son. The fund supports graduate students in the MSE department.

Belgacem Haba, arguably the star of the MSE soccer team of the late 1980s. (Bel Haba)

Chapter 14

International Students

International students have been a part of the Mining and Metallurgy Department since its founding in 1919. The first person to receive a graduate degree from the department that year was the Australian Welton J. Crook. While the number of international students remained small during the department's first few decades, by the turn of the twenty-first century, nearly 50 percent of our graduate students were international students.

Turkish students were among the first international graduate students, but in time students came from around the world, including Canada, Mexico, Brazil, and especially China, Korea and India. The story of the department's Korean students reveals the strong support groups they created for each other at Stanford, and that they would maintain throughout their careers.

Turkish Students

In the early 1940s, the government of Turkey began to send students to Stanford to study in various engineering fields. Because mining and metallurgy was an important engineering discipline for Turkey, several students had entered the Mining Engineering Department by the mid-1940s. By 1945, 28 students from Turkey were studying engineering, eight in the Mining Engineering Department.

As described in Chapter 4, Nevzat A. Tiner and Mehmet H. Erten became the first students to receive Ph.D.s from the department, in 1945 and 1946, respectively. Amet Servet Duran would receive his Ph.D., in 1963, after starting his career as a teacher at Washington State University. Duran and two others had studied at the Missouri School of Mines and had been recommended to study for Stanford's Master of Science degree by one of their Missouri professors.

Another Turk, Ayşe Emel Geçkinli, became the first woman to receive a Ph.D. from the department, in 1973 (see Chapter 12).

By the time of this report, 13 people from Turkey have been granted Ph.D.s by the department. They are listed here along with the year they received their degree and the name of their doctoral advisor:

Nevzat A. Tiner, 1945, O. Cutler Shepard

Mehmet Hayri Erten, 1946, O. Cutler Shepard

Amet Servet Duran, 1963, Robert A. Huggins

Tarık Ömer Oğurtanı, 1965, Robert A. Huggins

Mario Gomez, David Stevenson and Tarık Ömer Oğurtanı when Stevenson was on sabbatical leave in Europe and was lecturing at the Middle East Technical University in Ankara, Turkey, 1970. (MSE Dept.)

Ayşe Emel Geçkinli, 1973, Craig R. Barrett

Necmi Bilir, 1974, Theodore H. Geballe

Salim Çıracı, 1974, Walter Harrison

M. Turgut Gür, 1976, Robert A. Huggins

Eyüp Sabri Kayalı, 1976, Oleg D. Sherby

Mustafa Lütfi Öveçoğlu, 1987, William D. Nix

Sinan Cengiz Özkan, 1997, William D. Nix

Nevran Ozguven, 2007, Paul C. McIntyre

Kemal Levi, 2009, Reinhold H. Dauskardt

Korean Students

The MSE Department has had a special relationship with students from the Republic of Korea for half a century. In the 1960s and early 1970s, it was very rare for Korean students to study abroad, and only a handful had come to Stanford by then. But in the late 1960s, Ho-Bin Im and Jae Young Lee became the first Korean students to come to the campus to study materials.

Im studied with Richard Bube and worked on electron radiation damage in the form of traps for electronic holes in cadmium selenide (CdSe). His work was later found to be relevant to the properties of CdSe/CdTe (cadmium telluride) solar cells. Lee studied with Norman N. D. A. Parlee, in the Applied Earth Sciences Department, which was closely aligned with the Materials Science Department at that time. Parlee was known for his studies of liquid metal processes as related to steel making. Im received his Ph.D. in 1970 and Lee received his in 1971. They both returned to Korea to join the faculty at the newly established Korea Advanced Institute of Science and Technology (KAIST). KAIST had a strong connection to Stanford from its beginning as Frederick E. Terman, the former dean of engineering who had recently retired as university provost, led the international advisory committee that recommended its founding in 1970. Terman continued to act as an advisor to KAIST during the 1970s.

Im and Lee contributed to making KAIST one of today's leading research universities in the field of materials science and engineering, and both were pioneers in developing strong programs of materials education and research in Korea. They, and those who followed them, essentially created the Korean academic field of materials science and engineering. The examples they set soon led other Korean students to come

Korean MSE grads at the graduation ceremony, 2006. Left to right: Unoh Kwon, Seung-Min Han, Kang-Ill Seo, Yong-Won Lee, Dong-Woon Shin, Jungyup Kim, Dong-Ick Lee, Dae-Yong Kim, Chang-Beum Lee. (Seung-Min Han)

to Stanford's MSE Department for their Ph.D.s. At the time of this writing, 81 Koreans have been awarded Ph.D.s by the MSE Department, fully 10 percent of all Ph.D.s granted in the history of the department.

Space does not allow for a full report on their success in both industry and academia, but some examples follow: Kyung-Soo Ahn, who finished in 1982, later became the chairman of Noru Paint. Dong-Wha Keum, who finished in 1984, joined the Korea Institute of Science and Technology (KIST, not to be confused with KAIST) and later became director of KIST. Keum is currently serving as the president of KIST in Vietnam, where he is helping to develop the field of materials science and engineering.

Of the department's 81 Korean Ph.D. graduates, 33 have become professors in Korea and in the United States. Many of those who returned to Korea became professors at top Korean research universities, such as Seoul National University, KAIST, Yonsei University, Korea University, Hanyang University, and Sungkyunkwan University. In addition, many have served in important leadership positions, as exemplified by Ji-Beom Yoo, a vice president of Sunkyunkwan University. Many of those who joined industries in Korea rose to key positions, such as Ho-Kyu Kang, who became vice president of Samsung Electronics; Seok-Hee Lee, who became vice chairman and president/ chief operating officer of SK Hynix; and Jae-Won Choi, who became the executive vice chairman of SK Group.

The Korean students who studied in the department developed a wonderful tradition of caring for each other as if they were members of the same family. That surely also contributed to the success they had later in their careers. For many of the earlier students, it was their first time in the United States. Current students would pick-up newly arrived students at the airport and help them settle in. This often involved purchasing household items, as well as computers, cars, and other things that are essential for graduate student living in the States.

As the number of Korean students grew, a Korean MSE graduate student organization (called KMSE) was created to formalize these

KMSE Fall Picnic at Half Moon Bay, 2004. (Seung-Min Han)

traditions. A fall picnic for Korean students was one of the first events organized through KMSE. It was scheduled each year close to the start of the fall quarter so that the new students could be introduced to the KMSE group. Newcomers then learned about life as a Stanford student, and heard about the culture of each research group that the senior students had joined. The senior students also provided advice about choosing a research group and often advised new students on any openings that might be available in their research labs. The KMSE group also organized monthly research talks to learn about each other's research areas and to promote collaborations when possible. Each year, one senior was elected as

Korean Quals Party, 2005. (Seung-Min Han)

Byung-Tae Ahn (Ph.D. 1989) and Robert Huggins at KAIST in 2007. Ahn received his M.S. degree at KAIST working with Ho-Bin Im, the first Korean to receive a Ph.D. from the Stanford MSE Department. (Seung-Min Han)

president of KMSE and that person held considerable power over KMSE affairs.

The senior KMSE students were also active in helping the more junior students study for the qualifying exams by hearing their research presentations and asking questions during "mock quals." Some tips about the best way to approach questions were given, as was advice on how to improve a student's English communication skills. Studying for the oral qualifying exam was an intense day and night affair, especially for Korean students who had come directly to Stanford from Korea. The results of each exam were awaited with great anticipation. While a student was taking the exam, the senior students would wait outside the Peterson Laboratory, presumably for the good news but also for the party that would follow—if the student passed the exam. The student passing the exam was expected to host and pay for a dinner that evening for the entire KMSE group to celebrate the occasion and to thank each of the seniors for their help. The excitement of passing the exam seems to have overshadowed the burden of the large cost of the celebration.

KMSE also organized social events for the Korean students, such as the Friday night tennis matches. All Korean students were welcome to join in, even those with no tennis skills. The senior students were happy to teach those without prior tennis experience, much like the other ways in which they were helping their juniors. After most of the events, students would go to the Oasis (a popular pub in Menlo Park) for some beer. Once every year, Samsung would sponsor a tennis tournament, for both men and women, pitting the Korean students at Stanford against those at U.C. Berkeley.

The KMSE was like a home away from home where students cared for one another. On

William Nix at KAIST, 2015. Left to right: Doek-kee Kim, Seung-Min Han, Ki-Soo Kim, Nix, Yong-Suk Kim and Dong-Wha Keum. (Seung-Min Han)

Paul McIntyre at KAIST, 2017. (Seung-Min Han)

traditional Korean holidays, such as Chuseok (Korean thanksgiving) and the Lunar New Year, KMSE graduate students would get together to celebrate. In addition, many students got married during their time at Stanford, which provided additional opportunities for celebration. The alumni of KMSE have kept in close touch with each other through a KMSE alumni association, which meets annually for an end-of-year celebration. In addition to reminiscing about their time as seniors in Stanford's KMSE family, they continue their mentoring work by looking out for their juniors who are starting in their careers.

The Korean students who returned to Korea have maintained strong links with the MSE Department, primarily through invitations to MSE faculty to give lectures in Korea and to provide advice on university organization and direction. Robert Sinclair, who has had many Korean students, visited Korea often to give lectures at both Yonsei University and Seoul National University. In 2015, Robert Huggins visited the MSE Department at KAIST to give a series of lectures on batteries; he also served as a department advisor during that visit. William Nix and Paul McIntyre have also visited KAIST and given lectures there, in 2015 and 2017, respectively. The connections between the MSE Department and Korean materials science and engineering community made over a half century remain active to this day.

Chapter 15

Broadening the Faculty in the 1970s

In the early 1970s, Arthur Bienenstock recruited Clayton Bates, a rising young African American scientist at Varian Associates, whom he had known when they were both at Brooklyn Polytechnic Institute and Harvard. Bates went on to establish his own successful teaching and research program in photodetectors at Stanford, but he also took on a special role for decades as mentor and role model for black students in science and engineering at Stanford. The early 1970s also saw the departure of Craig Barrett to Intel, creating a vacancy in the area of electron microscopy, a subject already recognized as being central to the field. This brought the Liverpool-born Robert Sinclair to the faculty, and led to the creation, for the first time, of a strong program in transmission electron microscopy.

Clayton W. Bates, Jr ~ Bringing Photosensitive Materials to the Department

Clay Bates arrived at Stanford in 1972 after spending the better part of a decade in industry. As a specialist on photosensitive materials and devices, he had been one of Varian Associates' principal creators of an x-ray intensifier tube used in diagnostic radiology, a device that became standard equipment in hospitals throughout the world. Arthur Bienenstock, who had recently joined the Stanford faculty to work in the area of x-ray science, soon learned of Bates' work at nearby Varian (located in the Stanford Research Park). Bienenstock and Bates had known each other as graduate students and had collaborated on a study using electron spectroscopy for chemical analysis (ESCA) on

Clayton W. Bates, Jr., 1970s. (Clayton Bates)

Clayton W. Bates, Jr. with his daughter.
(Clayton Bates)

equipment at Varian to understand the properties of amorphous materials.

Bates was appointed as an associate professor with a joint appointment in electrical engineering. He was the first African American to hold a tenure track appointment in Stanford's School of Engineering, and also the first African American to earn tenure in the school.

On arriving, Bates immediately offered graduate courses in subjects close to his area of research: MSE 224, "Physical Properties of Disordered Materials;" MSE 233, "Quantum Theory of Energy States in Solids;" and MSE 248, "Photoelectronic Materials and Devices Laboratory," the last of which was cross-listed in Electrical Engineering.

Bates built a vibrant research group at Stanford that came to include about 10 students and post-docs. He and his students used x-ray photoelectron spectroscopy to study the structure and properties of a wide variety of materials. They also developed a spray pyrolysis and heat-treating method for creating copper indium selenide for various photovoltaic applications. Later in his career, Bates received patents for the development of composite photosensitive materials containing particles of an electrically conducting material embedded in and dispersed through a semiconducting matrix, which was transparent to the wavelengths of interest. His patents were based in part on a theoretical paper he wrote on photoexcited electrons in small metal particles with his graduate student, Quark Chen, in the mid-1980s and published in *Physical Review Letters*.

Clayton Wilson Bates, Jr. was born on September 5, 1932 in New York City, where he attended elementary school at New York Public School 119 and middle school at New York Junior High School 43. Having excelled at school, Bates next attended the prestigious all-boys Brooklyn Technical High School, graduating in 1950 as fifth in a class of 600, although not without some setbacks along the way. He reports that he failed the entrance exam the first time he took it, while his best friend passed. Hurt by this failure, he was determined to persevere. He passed the entrance exam the second time and later graduated well ahead of his best friend. For Bates, the lesson was clear: failures in life should be considered temporary setbacks to be overcome by persistence and effort.

It was in high school that Bates also first demonstrated his athletic prowess as a member of the baseball, basketball, and track teams, talents he brought with him to Stanford. He had enjoyed playing baseball and basketball as a young boy in his Harlem neighborhood, where he lived with his mother and older sister. (His parents had divorced when he was young.)

Bates has said that he could remember being interested in science and engineering from a very early age. On graduating from high school, he attended Manhattan College on a full academic scholarship with an initial plan to study aeronautical engineering, a field related to his boyhood dream of becoming a pilot. He started out as a civil engineering major but gravitated to electrical engineering and received his B.S. degree in that subject in 1954. He went on to earn his M.S. degree in electrical engineering from the Polytechnic Institute of Brooklyn. With the support of a fellowship, he then attended Harvard University, where he earned his second M.S. degree in electrical engineering in 1960. He continued his graduate studies at Washington University in St. Louis, Missouri, where he received his Ph.D. degree in physics in 1966, working with Edwin T. Jaynes.

After college, Bates worked briefly for several different companies in the areas of circuit design and solid-state physics before joining Varian Associates in late 1966 as a senior engineer to work on the x-ray intensifier tube. He had been recommended by E.T. Jaynes, who was at the time consulting with Varian Associates. It was during his time at Varian Associates that Bates had a chance to try out teaching at the university level. On sabbatical leave from Varian, he served as a Reader at Imperial College of Science and Technology of London in Kensington. He had been granted the sabbatical leave after only two years at the company, and ahead of some of his envious colleagues, as a reward for the important work he had done on the x-ray intensifier tube. That sabbatical later influenced his decision to accept the appointment at Stanford.

Clayton W. Bates, Jr. on one of his modeling gigs. (Clayton Bates)

In addition to his teaching and research, Bates was very influential in supporting African American students at Stanford. In 1973, soon after joining the faculty, he co-founded of the Society of Black Scientists and Engineers (SBSE), an organization "dedicated to fulfilling the mission of the National Black Society for Engineers (NBSE): to increase the number of Black engineers and scientists who excel academically, succeed professionally, and positively impact the community."

Bates pointed out in a 2004 *HistoryMakers* oral history interview that black students rarely have engineers in their families, and do not get the same career information as white students. Bates served as an advisor, formal and informal, to the increasing number of black students studying science and engineering at Stanford. He and his wife, Priscilla, also were resident fellows for three years, first at Roble Hall and then at Lagunita's East Court, where they guided the Black Theme House. Having grown up poor in Harlem—his mother, as a single parent, worked as a domestic to support her small family—he knew the obstacles that minority students face in life as well as at Stanford. He was ideally suited to advise students who faced problems similar to those he had overcome. Never one to offer a soft path or to mollycoddle, Bates is remembered for the high standards he set for his advisees.

Clayton and Priscilla Bates made a very handsome couple. Priscilla had worked as an actress when young and had the idea of getting into modeling. She brought Bates with her to see a photographer in San Francisco to have promotional photos taken. While there, the photographer asked if Bates would sit for photographs as well. Both sets of photos were submitted to a major advertising agency, along with those with other hopefuls. Bates was the only one selected for future modeling work, an unexpected outcome that probably did not go over well with Priscilla. Bates worked sporadically as a model for several years, appearing on the covers of magazines as a riverboat gambler, a sportsman, a guitarist, a basketball player, and a three-piece suit executive.

Throughout his career, Bates remained interested in minority science education and in the need for African Americans to be more involved in science policy. Committed to increasing the number of African Americans in STEM fields, Bates turned to improving scientific research more generally at historically black colleges and universities. In 1994, when he became an emeritus professor at Stanford, he accepted a faculty appointment in the College of Engineering, Architecture, and Computer Sciences at Howard University, the historically black university in Washington D.C. In his first assignment there, he established an interdisciplinary graduate program in materials science and engineering, the first of its kind in a historically black university. Two years later, in 1996, he was appointed associate dean for graduate education and research; he used his experience with the materials science and engineering program as a model to establish a broader set of interdisciplinary graduate programs at Howard. He also held an appointment as Professor of Electrical Engineering at Howard University. He brought a unique perspective to Howard University, with his own background, but also having worked at a predominantly white campus before moving to a predominantly black campus, with their inherent cultural differences.

Clayton W. Bates, Jr. (Clayton Bates)

In addition to a number of awards for his scientific work (including the "Distinguished Engineer" award from the national engineering society Tau Beta Pi, in 1977, for his work in developing medical diagnostic devices that use low x-ray doses) he received, in 1987, the Black Engineer of the Year award for Promotion of Higher Education.

Bates retired from Howard University in 2013 after 19 years there. He and Priscilla returned to live on the Stanford campus, where he continues to serve as Professor of Materials Science and Engineering emeritus.

Robert Sinclair ~ High Resolution Transmission Electron Microscopy

Robert (Bob) Sinclair was born in Liverpool, England, in 1947, and grew up there during Britain's difficult post-World War II years. Liverpool was a working-class port city, and the conditions at that time were austere. However, it was also a time of significant social change, most notably for Sinclair, with increasing educational opportunities for students from non-elite family backgrounds. In due course, he was able to pass the entrance examination to grammar school (St. Anselm's College, Birkenhead) and to university (Jesus College, Cambridge University). At Cambridge, he studied natural sciences, with an eventual B.A. degree (1968) in materials science, which was a new discipline in the U.K. in the mid-1960's. Sinclair stayed in the Department of Metallurgy and Materials Science for his Ph.D. (1972), being co-advised by John Leake and Brian Ralph, whose respective specialties were X-ray diffraction and field-ion microscopy. His thesis concerned spinodal decomposition of dilute nickel-titanium alloys, and provided some of the first direct evidence associated with the local compositional changes occurring during this type of phase transformation.

After completing his Ph.D. research in 1971, Sinclair took up his first post-doctoral position in the Crystallography Laboratory at the University of Newcastle-upon-Tyne, in the north east of England. His advisor there was Professor

K.H. Jack, a longtime expert in nitridation of alloys and development of silicon nitride ceramics. Sinclair's project focused on the former, extending Jack's ideas (substitutional-interstitial clustering) to understanding the mechanisms of the internal oxidation of alloys. It was during this period that he developed an interest in the applications of transmission electron microscopy (TEM), bridging the gap between his experience with bulk X-ray diffraction and with atomic-level field-ion microscopy.

Robert Sinclair, 2014. (Stanford News Service)

After about eighteen months in Newcastle, Sinclair took up the opportunity of a second postdoctoral position at the University of California, Berkeley, with Professor G. Thomas, perhaps the most influential practitioner of electron microscopy applications in materials science. The research project, funded by the National Science Foundation, was to study the use of the critical voltage effect in high voltage electron microscopy in ordering and ordered alloys. However, Sinclair's arrival in Berkeley in April 1973 coincided with the installation of a new transmission electron microscope (TEM), which was then capable of resolving the atomic planes in alloys at 2.0 to 2.5 Å resolution. Thomas agreed with Sinclair's suggestion that they try the newly emerging field of "high resolution electron microscopy" to study ordering before embarking on the higher voltage, but lower resolution, work. With the facilities available at the Lawrence Berkeley Laboratory, he was able to produce successful images of alloy ordering phenomena, resulting in the first publication from Berkeley on "lattice imaging." This led to further work on other metal phase transformations such as spinodal decomposition, precipitation, martensitic transformations, short-range order, etc. Sinclair's work at Berkeley led to his being selected to receive the Robert Lansing Hardy award in 1976 from the Metallurgical Society of AIME and the Burton Award of the Electron Microscopy Society of America (now MSA) in 1978.

At that time, in the mid-1970's, TEM was emerging as a much more diversified characterization instrument, as micro-chemical analysis, convergent beam electron diffraction, and electron energy loss spectroscopy were being developed, as well as high resolution lattice imaging. Stanford's MSE Department decided to search for a new faculty member in this field to replace the departing Craig Barrett, who was leaving to go to Intel. Sinclair was appointed to this position in 1977 after impressing the search committee and the MSE faculty with his novel images of lattice planes in crystals. That started a 40-plus year career at Stanford, during which Sinclair has remained dedicated to advanced electron microscopy as applied to materials research.

In his early years at Stanford, Sinclair devoted most of his research effort to setting up a TEM laboratory, and to generating interest in this approach more broadly at the institution. He was able to combine his funding from the

National Science Foundation (NSF) with equivalent funds from the Center for Materials Research (CMR) to purchase two new TEM's: one dedicated to high resolution imaging (which was his specialty) in MSE; and a second one mainly used for microanalysis, convergent beam diffraction, and diffraction contrast imaging in CMR. At that time, CMR appointed Dr. Ann Marshall to run their laboratory, which she has done with dedication, brilliance, and sensitivity to this day. Marshall's expertise complemented Sinclair's own interests, and their joint contributions over 40-plus years have established a central, powerful capability in materials characterization, which is one of the cornerstones of the MSE field.

Sinclair's early work explored new applications of advanced imaging in addition to alloy phase transformations, notably on deformation mechanisms in hard materials such as tungsten carbide and high-resolution imaging of semiconductors. It was his thriving collaboration with Richard Bube, examining cadmium telluride solar cell material interfaces, that led to a career-long interest in applying high-resolution TEM (HRTEM) to semiconductor materials, devices and integrated circuit technology. At that time, cadmium telluride (CdTe), silicon (Si), gallium arsenide (GaAs) etc. were more amenable to "lattice imaging" than were metals, as the former's lattice spacings were large enough for direct resolution by the contemporary HRTEM's with ca. 0.25nm point resolution. However, TEM specimen preparation was extremely difficult and several students sacrificed many hours of their time gradually perfecting their technique. Their contributions established a basic groundwork on which subsequent students could draw, and which put the Stanford group at a distinct advantage for cross-section specimen preparation. This later evolved to studies of metal multilayers, metal-semiconductor interfaces, magnetic thin films, and now more recently to nanomaterials for cancer research and to oxide interface structures.

When initially appointed, Sinclair was given the task of setting up several new courses, as well as contributing to the teaching of more general MSE classes. He established a new version of the first-year graduate course on "Atomic Arrangements in Solids," combining crystal structures of direct interest in MSE with basic crystallographic principles and diffraction (a course currently taught by Evan Reed). He took over one section of the undergraduate class

Ann Marshall, 2012 Marsh O'Neill Award Recipient, at the FEI Tecnai transmission electron microscope. (Stanford News Service)

E50, "Introduction to Materials Science," which he continues to teach; at one point, the class reached an enrollment of a few hundred students! He gained fame at Stanford for periodically holding that class on the lawn in front of the Peterson Laboratory on sunny Fridays during Spring Quarter. He also still teaches, in alternating years, the upper level graduate courses on "Transmission Electron Microscopy" and on "Nanocharacterization of Materials." The former has been naturally complemented by Marshall's hands-on laboratory course, which is offered every year. The latter realized a significant boost in importance when the Stanford Nanocharacterization Laboratory (SNL) was established in the early 2000's as a university-wide facility with multiple state-of-the-art instruments including TEM, SEM, XRD, XPS etc.

In recent years, Sinclair has become interested in teaching seminar-style classes which are of more general interest to undergraduates and which involve more discussion than chalkboard-based courses allow. This first started with his involvement as a faculty instructor in Stanford's Bing Overseas Studies Program (BOSP), teaching "Research in Japanese Companies" at Stanford's Kyoto Center in 1997. This course was based on his experiences carrying out semiconductor TEM research over several summers and one sabbatical quarter at Matsushita Electric Industrial Company in Osaka, as well as his involvement in a magnetic multilayer project with Bruce Clemens, funded at Stanford by the Kobe Steel Corporation. He brought this course back to Stanford when the introductory seminars program was established the following year, and he has taught ever since what has now evolved to be the "Japanese Companies and Japanese Society" course. This class continues to attract about twenty students each year from across the university.

He also has contributed to the BOSP with his "dream" course on "Soccer and English Society," which he has taught twice, in 2001 and 2010, at Stanford's Oxford Center. Most recently he taught a three-week overseas seminar in Cape Town in summer 2018 on "Soccer and Rugby in South Africa: a Racial Divide and Future Transformation." Through these opportunities, he has come to appreciate the equally valuable approach to learning afforded by such self-motivated, discussion-driven courses.

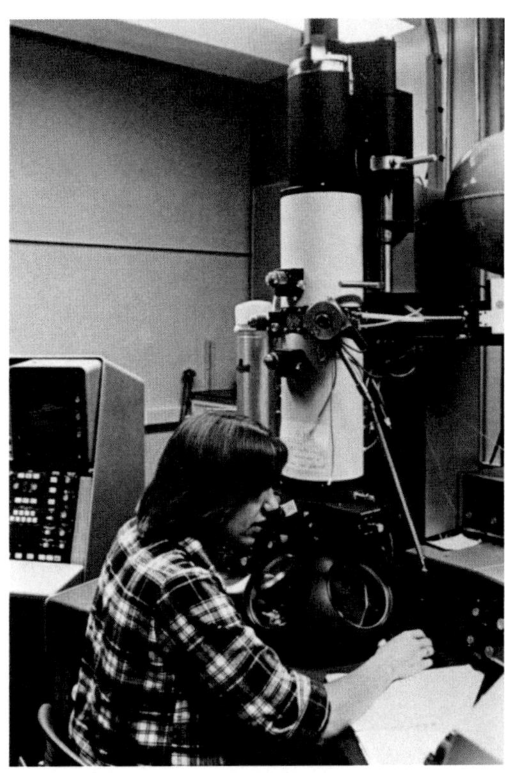

Ann Marshall, in 1980, at the transmission electron microscope, featured in a CMR brochure. (MSE Dept.)

As his research group was growing, and the point resolution of TEM's was passing the atomic dimension levels of 0.2nm, it was clear that a more up-to-date microscope was needed. The CMR invested in this purchase in 1984, which led to a completely new, more direct approach to advanced TEM examinations. At the same time, Sinclair was becoming interested in the possibilities of applying high resolution imaging at atomic resolution to *in situ* studies, and his group showed that this was indeed possible in a series of three *Nature* papers (one a cover article). This was at the time when *Nature* publications were quite rare within the MSE field, and indeed were not recognized broadly as being such useful contributions! (This paradox of course continues somewhat to this day.) His students expended significant effort into ensuring that the observations, and the information derived therefrom, were representative of bulk material behavior, oftentimes measuring activation energies of thermally activated processes *in situ* in the TEM versus ex situ in conventional annealing experiments.

This body of work applied to many materials systems over more than twenty years has led to recognition by several society awards such as the Distinguished Scientist Award (Physical Sciences) of the Microscopy Society of America (2009) and the David M. Turnbull Lectureship of the Materials Research Society (2012). Of course, nowadays, in the era of nanotechnology, the behavior of small or thin material has intrinsic value compared to the bulk, and the application of *in situ* TEM in MSE continues to grow year by year, and has expanded to gaseous and liquid reaction environments as well, inside the high vacuum required for a high resolution TEM.

Towards the latter part of his career, Sinclair was asked to assume some administrative responsibilities. Accordingly, he became the first director of the newly established Stanford Nanocharacterization Laboratory in 2002, and was able to gradually build it, with the support of the administration, faculty colleagues, and the research staff, into one of the premier materials characterization facilities at a U.S. research university. The 2009 acquisition, with financial support from five deans and the president of the university, of one of the first aberration-corrected environmental TEM's, now with 0.07nm resolution, at that time cemented Stanford's position as a world-class characterization institution. It also transformed the research capability of many faculty and students at Stanford, including his research group. In 2004, Sinclair became chair of the MSE Department, a position that he held for ten years, during which time he and associate chair Reiner Dauskardt were able to continue what Bruce Clemens, his predecessor, had set in motion. With support from the School of Engineering, the MSE department was able to hire a series of young, brilliant faculty, each expert in their own separate fields. These appointments led to wholly new research areas and ideas, and changed the direction and overall perception of the department. These new faculty members are now emerging as leaders of MSE, both within and outside the university. For two years Sinclair was also director of the Bing Overseas Studies Program (2010-12), and he is currently director of the Wallenberg Research Link, having succeeded Stig Hagström and Arthur Bienenstock in this capacity in 2017.

Sinclair's research has largely involved development of new, advanced TEM techniques with application across a broad spectrum of materials science. Thus, his group's early work on high resolution imaging of semiconductor interfaces in cross-section has led to the incorporation of this capability, albeit considerably more automated nowadays than in the past, in integrated circuit fabrication lines. The contributions to *in situ* high resolution TEM are perhaps the most well-known, but there have also been multi-Ph.D. efforts into the high resolution structure-property relationships in magnetic thin films and multilayers, into the reactions at material interfaces during annealing at elevated temperatures, and in a long time collaboration with Shan Wang (MSE/EE) and Sam Gambhir (Radiology) using high resolution and environmental TEM to characterize nanoparticles for possible medical applications such as early cancer detection and therapy. Sinclair remains convinced that TEM will have increasing applications in studying basic phenomena of direct relevance in materials research, from the truly atomic, to the nanometer and micrometer scales.

Chapter 16

Professors (Research)

By the mid-1950s, when federal support of research in universities was growing rapidly, Stanford responded aggressively with various schemes to increase the amount of sponsored research being conducted by the faculty. The availability of funding made it possible to persuade the university to provide more faculty billets to the School of Engineering, in return for a promise by the school that much of the funding for those billets would come from research grants and contracts. Thus, the practice of offsetting 50 percent of a professor's academic-year salary to grants and contracts was adopted, leading to a dramatic expansion in the number of faculty members in the School of Engineering.

During the 1960s, also fueled by availability of federal research funding, the university took another step along this path by creating a non-tenured research line in the professoriate, first called adjunct professor and later, in the early 1980s, professor (research). People appointed to this rank were expected to generate all of their salary and other funding needs through research grants and contracts. Nevertheless, they were held to the same standards that applied to tenure-line faculty, except that teaching was not required. With this teaching exception, professors (research) operated like tenure-line faculty members, with research laboratories of their own, including graduate students, post-docs, visitors, etc. Two people were appointed as professors (research) in the MSE department: Robert S. Feigelson and Alan K. Miller. The university also established a non-tenure teaching line in the professoriate, the rank of professor (teaching), where the responsibilities were purely teaching and funding was supplied by the department or school. No such appointments were made in the MSE department.

Robert S. Feigelson ~ Crystal Growth

Robert (Bob) Feigelson arrived at Stanford in the last few weeks of 1963 to work with Bob Huggins to both set-up a crystal growth laboratory in the ARPA-supported Center for Materials Research (CMR) and pursue a Ph.D. degree in materials science. After receiving his Ph.D. in 1974, he continued to serve as director of the crystal growth laboratory that he had established. It became a world-renowned crystal growth center within CMR, with strong ties to many of the Stanford faculty. He later joined the MSE faculty, first as an adjunct professor in 1978, and then as a professor (research) in 1983.

Bob was born and lived in Brooklyn, New York, until he was seven years old, when his family moved across the East River to midtown Manhattan. In Brooklyn, he says, he was a middleclass street kid constantly getting into trouble. But the move to Manhattan settled him down and there he began to pursue an academic career. After he graduated from New York's Stuyvesant High School in 1953, he left for the South, where he did his undergraduate work at Georgia Tech in Atlanta. It was a time of strong racial segregation and was a quite a shock to a young man from the North to experience segregated bathrooms and buses. He did, however, develop a strong Southern accent after six months in the Atlanta area, which appalled his family up North.

After graduating from Georgia Tech in 1957 with a B.S degree in ceramic engineering, he married his long-time New York City girlfriend, Vicki Zall, and shortly thereafter they were off to Fort Worth, Texas, where he had secured a job at Convair, an aircraft company and a division of the General Dynamics Corporation. At Convair, he was assigned to the materials processing group, which turned out to be more a materials procurement activity than a laboratory-based materials processing activity.
In addition, Convair had no immediate interest in using ceramic materials in their aircraft.

Without a laboratory or an interesting research project to pursue, and after a brutal company layoff that saw some his colleagues unceremoniously fired (here on Friday, gone on Monday), Feigelson made plans to return to school to get an advanced degree. He did so by searching for a job that would permit him to continue his education as a graduate student. That search resulted in his appointment, in 1958, to the staff of the Rodman Laboratories of the Watertown Arsenal in Watertown, Massachusetts, which permitted continuing education of its engineers, primarily at the Massachusetts Institute of Technology. He then had to sell the idea to the Ceramics Division of the Metallurgy Department at MIT, where he hoped to study. During an hour-long interview with the preeminent ceramist of the day, Professor F. H. Norton, he worked hard to convince Norton that he would be a hard-working student in their Master's program. Accepted into the Department in 1958, he graduated in 1961.

In the beginning, he worked with Professor Norton on a dielectric heating project, but later switched to Professor W. D. Kingery's group, where he worked on a new high temperature silicon-boron system. For his Master's thesis research, he synthesized and prepared dense polycrystalline structures of both silicon hexaboride and tetraboride and measured their physical and mechanical properties. That work led to his first conference presentation abroad, at a meeting of the British Ceramic Society in Kent, England, in 1961.

He left the Watertown Arsenal shortly after graduation for an exciting opportunity working at the newly formed Sperry Rand Research Center in Sudbury, Massachusetts. It was there that he started his crystal growth career, which was later to encompass the understanding of relationships between crystal growth processing parameters and the structure and properties of the crystals he produced. His first project was the growth of neodymium calcium tungsten oxide ($Nd:CaWO_4$) crystals for laser applications. He also grew, for the first time, crystals of the entire series of rare earth phosphates.

While his work at Sperry was interesting, he was one of only two on the technical staff who did not have a Ph.D. He found this situation unsatisfactory and started searching for a way to go back to school for an advanced degree. Around that time, Professor Robert L. White, who held a joint appointment in MSE and Electrical Engineering at Stanford, was spending some time at Sperry. He mentioned to Feigelson that he was looking for someone to grow ferrite crystals for his electron paramagnetic resonance studies. That led

White to suggest to Robert Huggins, then director of the newly created CMR at Stanford, that he bring Feigelson to Stanford to set up a crystal growing facility in the new center. Huggins did so, and further agreed that Feigelson could pursue a Ph.D. degree in MSE while working in CMR.

At Stanford, Feigelson was initially assigned a small laboratory in the courtyard of the Peterson Laboratory. The lab was equipped with a few small crystal-growing furnaces, including the first commercial silicon Czochralski crystal pulling system. While this furnace system was in use for many years, it was eventually placed on inactive duty for several decades, until it was refurbished and used during 2013-2016 for growing cesium iodide (CsI_2) crystals for nuclear detector applications. As expected, growing ferrite crystals for Professor White was a major priority for the new facility in the early 1960s, but Feigelson also worked on the growth of copper chloride (CuCl) crystals and expanded his research on rare earth phosphates to include the analogous arsenates and vanadates using lead pyroarsenate and pyrovanadate fluxes, respectively. Interestingly, in late 2017, Professor Ian Fischer (Applied Physics) found one of Feigelson's early publications on the rare earth vanadates while searching for information on thulium vanadate ($TmVO_4$) single crystals. Feigelson was able to find a suitable crystal grown in 1964, which Fischer was able to use successfully—almost 55 years later.

A major turning point in Feigelson's career came in 1966 when Professor Calvin Quate (Applied Physics) came back to Stanford from a visit to Bell Laboratories very excited about their work on lithium niobate ($LiNbO_3$) and other optical materials. He asked Feigelson if his group could grow single crystals of lithium niobate for Quate's own research projects. This led to an exceptionally successful interaction between the CMR crystal growth facility and the Applied Physics Department involving a large number

Robert S. Feigelson (Stanford News Service)

of faculty, including Professors Gordon Kino, John Shaw, Steven Harris, Richard Pantell, Bertram Auld, and Anthony Siegman, and over 25 students, including future Stanford Professors Robert Byer and Kenneth Oshman (Oshman was a co-founder of ROHM--later acquired by IBM). This interactive program involved optical parametric and Raman oscillators, acoustic devices such as the acoustic microscope and elastic surface waves. This important research collaboration was written up in a CMR booklet called "The Lithium Niobate Story: The Interrelationship Between Materials Synthesis and Device Research."

In the 1970's, with requests for bulk crystal growth dropping off because of changes in funding, Feigelson learned that John Shaw in Applied Physics was interested in having single crystal fiber lasers for use in a glass fiber gyroscope system he was building. Feigelson jumped at the chance to try to grow these single crystal fibers, and also contacted Robert Byer to see if he might have some parallel interest in lithium niobate ($LiNbO_3$) fibers. After some quick calculations,

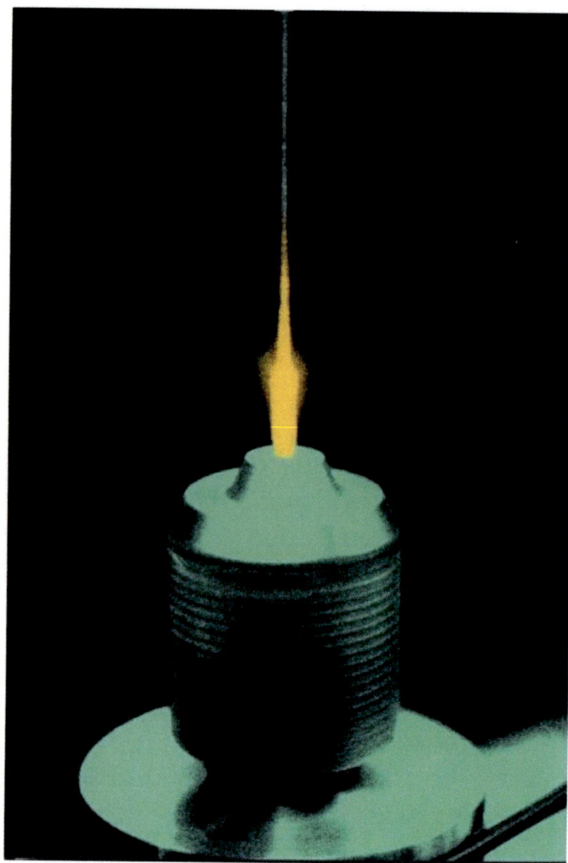

Pedestal growth of a fiber, Robert S. Feigelson. (Robert Feigelson)

Byer concluded that it would be very interesting to have $LiNbO_3$ fibers also, so a fiber growing research program was initiated. The research work in Feigelson's lab on the laser-heated pedestal fiber growth method became of great interest to both the international crystal growth and physics communities, and projects were funded by various agencies, including the National Security Agency (NSA). In collaboration with Theodore Geballe, the lab also grew barium calcium strontium cuprate ($Ba_2CaSrCu_2O_8$) superconducting fibers. This work, with graduate student David Fork, was published in Science with a cover photograph of a growing fiber.

Feigelson has been very involved in the international crystal growth and materials communities. In 1975, he organized the American Association for Crystal Growth's 3rd conference at Stanford. It was a very successful event that led to his becoming the organization's treasurer and later its 3rd president. He also was the editor of their newsletter for 10 years. He is currently a lifetime member of its executive committee.

In the early 1970s, Feigelson started a Northern California Crystal Growth organization, which included participants from many of the major materials companies and universities from San Jose to Berkeley. At the same time, he also started a Western Section of the American Association for Crystal Growth, which, after a few meetings in Southern California, settled on a permanent venue at Stanford's Fallen Leaf Lake Sierra Camp. These annual meetings have been ongoing since 1975.

In 1995, the International Organization for Crystal Growth awarded Feigelson their prestigious triennial Laudise Prize for contributions to experimental crystal growth. The ceremony was held in The Hague (Netherlands). Among other aspects of his career, he was cited for his work on fiber growth.

Robert S. Feigelson (Robert Feigelson)

In the course of his career, Feigelson received funding from over 25 different government agencies and industries. The main focus was on optical materials but other areas such as thermoelectrics, semiconductors, polymers and proteins were also investigated. In addition to growing bulk crystals, fibers and epitaxial thin films using Metal Organic Chemical Vapor Deposition (MOCVD), he also worked on preparing transparent ceramics. Feigelson has published almost 300 papers in refereed journals. In addition, he has edited some special issues for Elsevier, such as "50 Years of Progress in Crystal Growth," and written a chapter entitled "Crystal Growth for the Ages" in the *Handbook of Crystal Growth* (2nd Edition). He also has consulted for many local and international companies and founded a small company to manufacture crystal growth equipment. For several decades he has been one of the principal editors of the Journal of Crystal Growth.

Alan K. Miller ~ Modeling Mechanical Behavior and Thermoplastic Composites

Alan K. Miller served on the faculty in the Department of Materials Science and Engineering from 1977 to 1990, first as a non-billeted assistant professor and later as a professor (research). He had received his Ph.D. in MSE in 1975 for his development of MATMOD (MATerials MODel), a system of constitutive equations for treating many types of deformation behavior within a unified phenomenological framework. He and his students later engaged in research on thermoplastic-matrix, fiber-reinforced composites, which led to his subsequent career at Lockheed. Although not required of a professor (research), Miller taught occasional courses in the MSE Department, including "Life Prediction in Engineering Structures" and "Failure Analysis."

During his time on the faculty, he received a number of awards for his work, including

Alan K. Miller (Alan Miller)

the Sam Tour Award in 1986, which was presented to him and his student Amaneh Tasooji by the American Society for Testing and Materials (ASTM) for their work on corrosion testing methods. In 1989, he also received an award in the Ameritherm Research Grant Competition; the prize was a $20,000 induction-heating unit, which he used in his subsequent research. He had earlier received the John Dorn Memorial Award in Materials Science from the Department of Materials Science and Engineering at U.C. Berkeley for his Ph.D. dissertation work. He left Stanford in 1990 to accept a position at the Lockheed Palo Alto Research Labs, where he had a long and distinguished career in new methods for fabrication of polymer-matrix composites.

Alan Miller was born in Brooklyn, New York, on Dec. 28, 1945 of parents who were in the garment trade. He grew up in Manhattan's Stuyvesant Town and on Long Island (Cedarhurst) and attended local public schools, developing an early interest in science and technology. He received Lawrence High School's award as "outstanding science student in scholarship and experimentation." He attended Cornell University's College of Engineering, benefitting from work experience within its Co-op program in making the choice of his ultimate career specialization (the intersection of Mechanical Engineering and Materials). He graduated with distinction in mechanical engineering in 1967.

Miller began his graduate studies in the Mechanical Engineering Department at Stanford in the fall of 1967, working with Professor Henry Fuchs and taking several Materials Science and Engineering courses. He received his M.S. degree in mechanical engineering in 1968, not long after meeting his future wife, Leta, a Stanford undergraduate. Leta would later have a long and productive career as Professor of Music at U.C. Santa Cruz, before becoming *emerita* in 2016.

With the Vietnam War escalating, 1968 was the year that U.S. Army draft deferments for graduate students were eliminated. With an occupational deferment as an engineer, Miller went to work for Pratt & Whitney Aircraft (PWA) in East Hartford, Connecticut. He was especially attracted by their Advanced Concepts group, which estimated the properties of materials likely to be available five to 10 years into the future, and then designed advanced gas turbine engines around those properties. At PWA, he also encountered the subject that he later pursued for his Ph.D. dissertation, namely the development of unified constitutive equations to better characterize the creep and cyclic plasticity response of high-temperature materials. After working at PWA for four years, he decided he would like to return to Stanford to work for a Ph.D., this time in the MSE Department. His former advisor in ME, Henry Fuchs, suggested that he work with Oleg Sherby. This began a wonderful 45-year association with Sherby, with Miller as a student, faculty colleague, and good friend.

Oleg Sherby and Alan Miller (Alan Miller)

In 1974, as Miller was nearing the end of his Ph.D. studies at Stanford, his wife was still in the middle of her Ph.D. program (music) at Stanford. He needed to look for a position in the Stanford area. To get acquainted with a potential local Bay Area employer, he took a summer job at the newly founded Electric Power Research Institute (EPRI). There he met Terry Oldberg in the Nuclear Power Division of EPRI, to whom he mentioned his Ph.D. dissertation on constitutive equations for metals and alloys. Oldberg expressed a need for constitutive equations to model the deformation of Zircaloy cladding used in nuclear reactor fuel rods. This led to the idea that Miller might work for EPRI to supply that need as a regular employee, after completing his Ph.D. But by Spring 1975, EPRI had undergone so much expansion that they were now forced to institute a hiring freeze. Instead of becoming an EPRI employee, Miller stayed on in the MSE Department as a research associate with Sherby, working on Zircaloy -- with the support of an EPRI contact.

As the initial work on constitutive equations for Zircaloy proceeded, Sherby suggested that the approach might also appeal to the Office of Basic Energy Sciences in the Department of Energy (then the Energy Research and Development Administration -- ERDA). Their proposal received a positive response from the agency, but at Stanford, a research associate could not be a principal investigator. Nor could Sherby accept the grant because he already had three

U.S. government programs. Sherby approached William M. Kays, then the Dean of Engineering at Stanford, looking for a solution to the dilemma. Kays suggested that the MSE Department put Miller on the faculty as a non-billeted (salary coming from research) assistant professor, a position that came with the ability to be both a P.I. and a principal supervisor of Ph.D. students. MSE did so. Eventually Miller progressed to non-billeted associate professor and then to professor (research).

Those initial years centered on improving and extending the MATMOD constitutive equations to treat many types of deformation behavior (and their interactions) within a unified phenomenological framework. It then migrated into predicting fatigue crack initiation and growth, stress corrosion crack initiation and growth, and sheet metal forming. Altogether, twelve students received their Ph.D.s in this work.

Miller was planning to extend his modeling to "die-less forging" of metals when Arden Bement, then a former director of DARPA's materials science office, suggested that he consider thermoplastic-matrix composites instead. These materials were just then emerging as fiber-reinforced polymeric materials that could be heated to elevated temperatures, formed to shape, and cooled back down to a rigid state. Thus "die-less forging of metals" soon became "die-less forming of composites." Five more students would receive their Ph.D.s working in thermoplastic-matrix composites.

When Miller joined the MSE department in 1975, the department was dealing with a broad mix of subjects that included materials for both structural and microelectronic applications. But by the 1980s, more and more attention was being focused on materials science issues related to the needs of Silicon Valley, and less and less attention was being paid to the mechanical behavior of structural materials, especially their engineering aspects. Being inherently a mechanical-devices-engineering type, Miller found himself increasingly out of the intellectual center of gravity of the department. When, in 1989, Purdue University advertised for a tenure-track faculty member specializing in fiber composites fabrication, Miller felt he had to apply. He was offered the Purdue appointment, as well as offers of endowed chairs from both the University of Wisconsin and USC. But his wife was up for tenure at UC Santa Cruz, and for other family reasons, he decided to look for a suitable opportunity within commuting distance of Santa Cruz. That led ultimately to an offer, in 1990, to join the Lockheed Palo Alto Research Labs to work in the area of composites. Before accepting Lockheed's offer, he consulted with Stig Hagström, then MSE chair, who offered some wise counsel. Hagström's analysis was that: (1) at his core, Miller was an engineer, and (2) "in engineering, the action is in industry." (Miller stayed on part-time at Stanford through 1993 to complete the guidance of his Ph.D. students.)

His subsequent 24 years at Lockheed focused entirely on composites. Miller's thermoplastic composites experience got him involved in an early Lockheed project on *in-situ* consolidation of thermoplastics for submarine structures that were "too big to autoclave." He brought his student, Rob Van den Niewenhuizen, over to Lockheed to do his dissertation research in that area. Miller later worked on a diverse array of projects involving both thermoplastic and thermoset-matrix polymer composites. One of the first was on the construction of light-weight composite highway bridges which could be prefabricated offsite, transported to site, and quickly assembled in a few days. Their decks were later commercialized as "Duraspan," which was used in dozens of bridges whose concrete decks had deteriorated. Later in his career he took a leading role in Lockheed's work extending Vacuum Assisted Resin Transfer Molding (VARTM), which had been developed in the boat-building industry, to aerospace structures. Miller investigated that process with lab experiments and by creating an Excel-based computer flow model that was used by Lockheed for the remainder of the time that Miller was with the company.

The highlight and the capstone of Miller's career at Lockheed involved his technical leadership within Lockheed's Ocean Thermal Energy Conversion (OTEC) project, which was part of Lockheed's diversification into alternative energy. Miller welcomed the project inasmuch as he was looking for a way to apply his engineering skills to the mitigation of global warming. The OTEC process requires a Cold Water Pipe (CWP) about 1000m long to reach down from the ocean surface to the cold water which serves as the coolant in the OTEC thermal cycle. It also needs to about 10m in diameter to handle the huge water flow of a commercially viable plant.

From earlier work, it had been guessed that the CWP would be a fiber composite. Miller, with his expertise in fabricating large composite structures, was asked to lead the CWP project. Through that project, Miller developed what became a life-long interest in the environment and how technology can be used to mitigate global warming.

Miller retired from Lockheed-Martin in 2014, but he continued working in the composites fabrication area for two more years at Specialized Bicycles. He maintained his interest in mitigating global warming by auditing courses in this field at UC Santa Cruz, by his own investigations (available for perusal at https://www.cool-it-earth.com/), and by working with Citizens Climate Lobby. He and his wife, Leta, continued to live in Santa Cruz, where they celebrated their 50th wedding anniversary in 2019.

Sadly, Alan Miller died on July 29, 2019, from a brain injury sustained in falls the previous year.

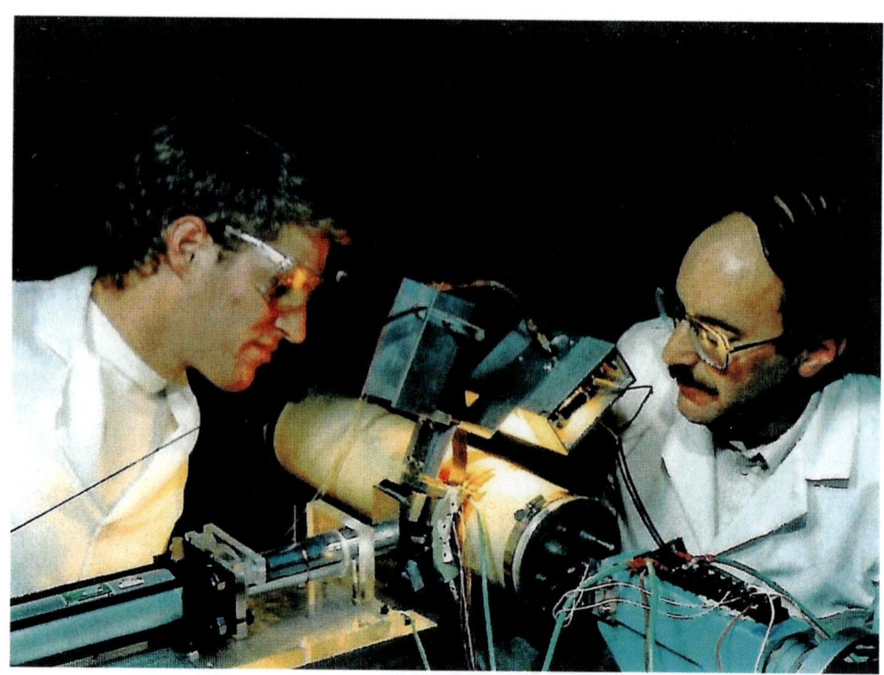

Rob Van den Niewenhuizen and Alan Miller working on the processing of a thermoplastic composite for a marine application. (Alan Miller)

Chapter 17

Joint Faculty Appointments

By 1962, Richard Bube had been appointed to the faculty with a primary appointment in Materials Science and a joint appointment in Electrical Engineering, but no faculty members with primary appointments in other departments had yet held a joint appointment in Materials Science. In an effort to further strengthen the faculty associated with the newly formed Center for Materials Research (CMR), particularly in areas related to electronic, magnetic and optical properties of materials, Center director Robert Huggins negotiated with the School of Engineering to make three additional faculty appointments in these areas in the departments of Materials Science and Electrical Engineering. Thus, Robert L. White (1962) and Kenneth A. Wickersheim were appointed to the Materials Science faculty with joint appointments in Electrical Engineering and William E. Spicer (1963) was appointed in Electrical Engineering with a joint appointment in Materials Science. While Wickersheim returned to industry not long after his appointment, White, a future chair of Electrical Engineering, and Spicer, a co-founder of the Stanford Synchrotron Radiation Laboratory, went on to productive careers at Stanford.

A few years later, Theodore Geballe came to Stanford (1968) as a member of the Applied Physics faculty with a joint appointment in Materials Science. He, too, would have an enormous impact on materials research at Stanford and on the training of many prominent people in materials science. Two later joint appointments with Materials Science are noteworthy: Friedrich "Fritz" Prinz (1994), of Mechanical Engineering, and Andrew Spakowitz (2014), of Chemical Engineering.

William E. Spicer ~ Photoemission Spectroscopy and Photocathodes

William E. Spicer joined the faculty as an associate professor of electrical engineering in 1963 with a joint appointment in materials science. He began by teaching a course on "Elementary Electromagnetic Theory" in Electrical Engineering. The following year, he started teaching his first cross-listed EE-MSE course, "Optical Properties of Solids." That course was in line with the work he had done at RCA, and with his developing Stanford reputation in the field of photoemission spectroscopy.

Spicer had worked for seven years as a research physicist at the RCA Research Laboratories in Princeton, New Jersey. His careful study of photocathodes (the "eyes" of the original TV cameras) gave him a deep understanding of the photoemission process, which remained the focus of his career. As

William E. Spicer (American Institute of Physics)

explained in his obituary in *Stanford Report* (2004), "in photoemission, light strikes a material and the interaction causes electrons to emerge. Data about the energy of the emerging electrons combined with knowledge of how light excited the electrons gives insight into the electronic structure of the solid. This knowledge found application in TV pictures and "electric eyes."

Working as a consultant to nearby Varian Associates, Spicer invented an improved x-ray image intensifier subsequently used in Varian medical devices. By intensifying image brightness by a factor of 10,000, the device let medical professionals see such objects as kidney stones for the first time. Its televised x-ray images also allowed them to examine movement in real time via, for example, the motion of blood vessels in a heart prior to a bypass operation.

It was Spicer, who first suggested that a port be built at the Stanford Positron Electron Accelerating Ring (SPEAR) to take advantage of its by-product synchrotron radiation for solid-state studies. Spicer later served as acting director, deputy director, and consulting director of the resulting Stanford Synchrotron Radiation Project.

> Recognizing that synchrotron radiation could provide a superior excitation source, in 1968 Spicer wrote to SLAC Director W.K.H. Panofsky to explore the possibility of using radiation produced by the linear accelerator and later stored in a hollow ring for experiments in fields besides high-energy physics. As a result, and with the efforts of other professors and graduate students, the first port was added on SPEAR, which was put into operation for high-energy physics under the leadership of Burton Richter in the early 1970s. That and subsequent ports siphoned off beams of x-rays and ultraviolet light for use in studies of matter at the atomic and molecular scales, and paved the way for advances in condensed-matter physics, materials science, chemistry and biochemistry. (*Stanford Report*, 2004)

William Edward Spicer had been born in Baton Rouge, Louisiana, on September 7, 1929. He overcame not only the limited opportunities of poor local school system but both dyslexia and a speech defect to earn, at age 19, a bachelor's degree in physics from the College of William and Mary, in Virginia, along with election to Phi Beta Kappa. He went on to earn a second bachelor's degree in physics from MIT in 1951, and then his master's (1953) and doctoral (1955) degrees from the University of Missouri. A prolific author (the author of more than 700 publications, and one of the 50 most-cited authors between 1982 and 1997), Spicer became an expert in the electronic structure of solids, with an emphasis on surfaces and interfaces, and on synchrotron radiation photoemission studies of semi-conductors.

In 1981, he told I*ndustrial Research & Development magazine* that his "disabilities" may have been a key to his success:

> I have since learned from my wife, Diane, who is a reading specialist, that how one learns is dependent very much on how one's mind is 'wired up.' People like me have to learn to think in different patterns from the norm. Not until after I got my Ph.D. did I learn that my pattern of

thought is essentially pictorial, whereas for most scientists it is mathematical. So, I have a very different approach to problems. *(Spicer, 1981)*

Spicer was named the Stanford W. Ascherman Professor of Engineering at Stanford in 1978. In 1980, he won the Oliver E. Buckley Solid State Physics Prize from the American Physical Society for "effective development and application of photoelectron spectroscopy as an indispensable tool for study of bulk and surface electronic structure of solids," and the following year was chosen as Scientist of the Year by *Industrial Research & Development* magazine. In 1984, the American Vacuum Society gave him its senior research prize, the Medard W. Welch Award.

Spicer was an exceptional mentor to more than 100 doctoral students and postdoctoral fellows. In 2000, he was awarded the Lifetime Mentor Award by the American Association for the Advancement of Science (AAAS).

Electrical engineering professor William Spicer, at right, in 1965. (Stanford News Service)

Spicer died of heart failure while vacationing in London in 2004.

Robert L. White ~ Magnetic Storage Technology and Biomedical Applications

Robert L. White came to Stanford in 1962 with a joint faculty appointment in the Electrical Engineering and Materials Science departments after several years in industry. He initially taught "Basic Quantum Mechanics for Engineers" in the Electrical Engineering Department, but the course was later listed in Materials Science as well. He also taught MSE 248, "Magnetic Phenomena in Solids," which led naturally to his work in magnetic storage technology and, later, to the use of magnetic nanoparticles for biomedical applications. Late in his career he taught a course entitled "Engineering in Biology and Medicine," which reflected his early work on microfabricated devices for auditory cochlear prosthesis and his work on magnetic nanotags for biomolecular detection technology.

White served as chair of the Department of Electrical Engineering in the 1980s and was instrumental in building stronger links between the Electrical Engineering and Materials Science & Engineering Departments. When the department was rebuilding in the early 1990s, he made the case for a joint faculty appointment between EE and MSE in the broad area of information storage technology. He had co-founded the Center for Research on Information Storage Materials (CRISM) at Stanford in 1991 and was anxious to strengthen the faculty in that area. The search led to the appointment of Shan X. Wang, who would later transform CRISM into the Stanford Center for Magnetic Nanotechnology, with a focus on the use of magnetic technology in biomedical applications.

Robert L. White (the middle initial is used to distinguish him from Robert M. White, who also works in the magnetics area and is closely associated with the department) was born in Plainfield, New Jersey, on Valentine's Day in 1927. His father had a small company that manufactured metal parts. He started his secondary education at Plainfield High School in 1944, but his studies were interrupted by service in World War II. After serving in the U.S. Navy in 1945-46, he finished high school in 1949 at the age of 22. He enrolled at Columbia College in New York in the fall of 1949 to study physics and mathematics and graduated with a B.A. degree in three years. He studied physics and mathematics as a graduate student at Columbia and received an M.A. degree and a Ph.D., the latter in 1957.

After receiving his Ph.D., White worked for a time at the Hughes Research Laboratory in Malibu, California, before moving, in 1958, to Sylvania, a division of General Telephone and Electronics (GTE) located on the San Francisco Peninsula not far from Stanford University. He worked there until his 1963 appointment at Stanford. He consulted for both Lockheed and Ampex, both nearby, during his early years at Stanford.

White's early Stanford research involved studies of the magnetic properties of rare-earth orthoferrites, $HoFeO_3$ and $YFeO_3$, for possible device applications. His work on magnetic phenomena in materials was a precursor to his later career focus on magnetic storage technology and the use of magnetic technology in biomedical applications. By the late 1970s, White was also beginning to work in the area of electronics for biomedical applications. With support from the National Institute of Health, he and his students were making use of planar lithographic techniques to create microelectrode arrays to produce localized electrical stimulation of the auditory nerve. This was in connection with the development of an auditory cochlear prosthesis, an implantable device to restore sound perception to people with profound hearing impairment. This effort made use of facilities at the newly founded Center for Integrated Systems (CIS), which at the time was housed in the basement of the McCullough

Fabian Pease and Robert White in the Magnetics Lab, McCullough Building, in the mid-1960s, from the department's CMR brochure. (MSE Dept.)

Building. It became a major Stanford facility for research on materials and devices when it moved into the CIS building in 1985, and it later became the Stanford Nanofabrication Facility (SNF), which is now included in the Stanford branch of the National Nanotechnology Coordinating Infrastructure (NNCI). The NNCI at Stanford consists of four facilities, SNF and the comparably sized Stanford Nano Shared Facilities (SNSF), as well as two much smaller facilities in Earth Sciences: the Mineral Analysis Facility (MAF) and the Inductively Coupled Plasma Mass Spectrometry facility (ICP-MS) facility.

By the mid-1970s, White was assuming leadership roles in the Department of Electrical Engineering; he served as associate chair of the EE Department from 1976 to 1982, and then as chair from 1982 to 1987. Following his time as chair of EE, he took a leave of absence from Stanford to accept an appointment as director of The Exploratorium in San Francisco, hoping to expand and improve the museum using the development approaches that had been successful at Stanford. But he found it difficult to change the culture at the museum and returned to Stanford after three years. During his time at the Museum, he had retired from the regular faculty to become Professor Emeritus of EE and MSE.

On his return to Stanford, White recognized that, although the area around Stanford was already called Silicon Valley—because of the use of silicon in computer chips—a case could be made for calling it Information Storage Valley. Following the lead of IBM in San Jose, which had introduced the first random access disk storage system in the 1950s, companies all over the valley were making magnetic media hard and floppy disks and their associated drives. White saw the need for university research in that area and, with support from industry, created the Center for Research on Information Storage Materials (CRISM). Research on information storage became the focus of many faculty and students in the department in the 1990s and after.

Robert L. White (Stanford News Service)

The discovery of Giant magnetoresistance (GMR) in 1988 by Albert Fert of France and Peter Grünberg of Germany led to extremely sensitive magnetic sensors that transformed magnetic storage technology. (GMR is an effect in which an electrical resistance is very strongly affected by the presence of a magnetic field.) The detection of magnetized bits in magnetic media is made profoundly more sensitive through GMR. This groundbreaking work led to the Nobel Prize in Physics for Fert and Grünberg in 2007. It was soon after this development that Shan Wang joined the faculty and began working in the magnetics area with White and others. Wang recognized that the sensitivity of magnetic field detection provided by GMR sensors would allow tiny bits of magnetic material to be easily detected. If biomolecular species were bound to magnetic nanoparticles, GMR sensors could then be used to monitor very small concentrations of these species, an advance which led to molecular diagnostics such as cancer diagnosis and treatment monitoring. Through this development, CRISM was transformed and renamed the Stanford Center for Magnetic Nanotechnology (CMN). In addition, magnetoresistive random access memory (MRAM) based on tunneling magnetoresistance (TMR) became another focus of the CMN.

In 2005, White and Wang and others founded MagArray Inc. to develop ultra-sensitive immunoassay systems for detecting biomolecules or biomarkers labeled with magnetic nanoparticles. The company develops diagnostic tests for several cancers, as well as cardiovascular, infectious, and inflammatory diseases. More recently, the company has started to offer ligand binding assay services to drug development and pharmaceutical companies.

White has received much recognition for his work. He is a Fellow of both the American Physical Society and the Institute of Electrical and Electronics Engineers (IEEE). White also served as a Distinguished Lecturer for the IEEE Magnetics Society. He held a Fellowship from the John Simon Guggenheim Foundation on three occasions in the 1960s and 1970s and he was also a Fellow of the Christenson Foundation in 1986.

White lives with his wife, Phyllis, at the Vi at Palo Alto, after having sold his campus home following his retirement.

Theodore H. Geballe ~ Semiconductors, Superconductors, and Interdisciplinary Research

Theodore H. Geballe, known to all as "Ted," joined the Stanford faculty in 1968 as a professor in the Applied Physics Department with a joint appointment in Materials Science. He came directly from Bell Labs where he had spent the previous 16 years and where, he says, he "learned solid state physics from where much of it was created." He had arrived at Bell Labs just three years after the discovery of the transistor.

In his first work at Bell Labs, Geballe measured the thermal conductivity of single crystal silicon and found it to be higher by a factor of two than the literature value for silicon. He soon realized that the vastly improved purity and perfection of the silicon that had been created for the new transistor technology was responsible for the difference. That realization made an indelible impression on him as it showed that the "interactions between science and technology provide a two-way street for advancing condensed matter physics," something that he witnessed for the rest of his career. His finding that defects play a role in the physical properties of crystals also brought him into the field of materials science.

Theodore H. Geballe (T. Geballe)

Ted Geballe was born in San Francisco on January 20, 1920 and was schooled there though high school. In 1937, he graduated from Galileo High School, which Joe DiMaggio had attended before dropping out to play baseball. The students were much more excited about "Joltin' Joe" hitting homeruns over the fence than in Galileo's dropping balls from the Tower of Pisa.

Geballe's first lesson in chemistry was given to him by his Aunt Pauline, who was a high school teacher in Portland when he was in grade school. His Aunt told him: "Johnny was a chemistry student. Johnny is no more, for what he thought was H_2O was H_2SO_4." Her most famous student was Linus Pauling, when he was a 12-year-old, and they remained in contact throughout their lives. Pauling recalled how she demonstrated air pressure by boiling water in a can of maple syrup. She screwed the lid on the can and when the steam inside condensed the can collapsed. John Galt, another famous alumnus from Bell Labs, with whom Geballe became close friends, had also been a student of Aunt Pauline.

After high school, Geballe "went east" to college by taking a ferryboat across the bay to Berkeley (the new Bay Bridge opened that same year). At U.C. Berkeley, he studied chemistry, and after a failed attempt to measure the heat of evaporation of electrons from a metal, better known as its work function, he was introduced to Professor William Giauque, who would become his advisor. Giauque, a later winner of the Nobel Prize in chemistry for his work on the properties of matter at a temperature of absolute zero, was at that time interested in the general problem of how materials lose or retain entropy when they are cooled to their ground state at T = 0, using the Third Law of Thermodynamics. As a part of this work, he was measuring the heat capacity of FCC metals to see how well the Debye approximation worked. He used some prize money he had received to purchase pure gold and offered Geballe the opportunity to make the measurements. Geballe grew the gold into a single crystal following Bridgman's technique by melting it in a cylindrical graphite crucible and cooling it, starting from a narrow tip at the bottom, making it suitable for measuring in the Giauque lab. That later led to his graduate work with Giauque and also to their joint publications, one on the heat capacity and entropy of gold.

Students at Berkeley in the late 1930s were encouraged to take Reserve Officer's Training, so Geballe enrolled in a program with the Army Ordinance Department. He was called for active duty as a second lieutenant in 1941, just a week before he graduated with a B.S. degree in Chemistry. He served mainly in the South Pacific in Australia, New Guinea, and the Philippines for the duration of the war. He returned to Berkeley afterwards and enrolled in its graduate program in Chemistry, again under the supervision of Professor Giauque. He received his Ph.D. in Physical Chemistry in 1949, the same year that Giauque won the Nobel Prize in Chemistry.

Following his initial work on semiconductors at Bell Labs, Geballe had an opportunity to work on superconductors with Bernd Matthias, with whom he formed a 20-year close association that resulted in nearly 40 joint publications. His paper with Matthias and V.B. Compton on superconductivity, published in *Reviews of Modern Physics* in 1963, was his most highly cited paper from that period. That led ultimately to his continued interest in superconductivity and to his leading role in the Stanford effort on high-temperature superconductors, after the 1986 report by Bednorz and Müller of possible high-temperature superconductivity in the Ba-La-Cu-O system.

Geballe's views on his work at Bell Labs and the later decline of that laboratory, which occurred when the government broke up the Bell System and split it into two separate companies, are revealing and insightful. Initially his view was that monopolies are not necessarily bad. In his first few years at Bell Labs, he found that the place was open to scientists from other industrial labs, such as GE, IBM, RCA, and others, who were interested in developing in-house capabilities to make transistors and other solid-state devices. The openness was practiced to protect AT&T's monopoly in communications by demonstrating that they were not competing in other markets. When the government finally broke up AT&T, Geballe and many others felt that the action was "killing the goose that laid the golden egg."

In retrospect, he later concluded that he had been wrong. Because of the breakup, AT&T no longer supported fundamental research, with the consequence that leading Bell scientists left the lab and joined the faculties of major research universities. That, in turn, jump-started

Ted Geballe and his family's golden retriever Charlie in 2016. (T. Geballe)

academia's ability to not only do modern research but also to educate the next generation of students. Stanford's MSE department greatly benefitted from the appointments of other refugees from the industrial laboratories, including Richard Bube (RCA), William Tiller (Westinghouse), William Spicer (RCA), Robert White (Sylvania), Clay Bates (Varian Associates) and Stig Hagström (Xerox).

At Stanford, Geballe continued to study superconductivity and other physical properties of metals, intermetallic and layered compounds. His group profited from advances in thin-film deposition and characterization techniques that had developed rapidly in semiconductor technology. The deposition systems that he and his collaborators built up allowed them to extend phase diagrams to obtain metastable and amorphous phases, and to deposit multilayers, heterostructures, interfaces and tunnel junctions. One of these systems was the planar magnetron sputtering system that was developed by Doug Keith and Troy Barbee in the Center for Materials Research.

The more applied aspects of Geballe's passion for science led to very strong collaborations with others in MSE, associations that he has maintained for half a century. He is a pioneer of interdisciplinary science, and was practicing it long before it became fashionable. He has an appreciation of how one needs all of the physical sciences to come together to make major advances. He has always appreciated theorists, experimentalists, and applications engineers equally, and has been able to bring them together to address important problems. Indeed, his ability to join people together across disciplines and institutions is one of Geballe's unique gifts.

The most exciting collaborative time with others in MSE was during the heady days of the discovery of high temperature superconductivity. During this time, Geballe worked with a group of very eager young scientists who shuttled effortlessly, frequently, and rapidly between the Departments of MSE, Applied Physics, and Physics, as well as between the independent labs of CMR and the Ginzton Lab. The Ginzton Lab eventually morphed into the Spilker Lab, while CMR later became the Geballe Laboratory for Advanced Materials. It didn't matter that the vapor deposition lab in CMR was shared by all and sundry, or that the lab management was "student led." World-leading science was to be done, and Geballe's disciples would go to any lengths to share in his youthful and infectious excitement, aiming to fulfill both his and their dreams. They all felt part of Geballe's family when working with him and that feeling persists to the present. Some of the notable former students and postdocs from MSE who worked with Geballe are Chang-Beom Eom (now the Geballe Professor at the University of Wisconsin, Madison); Bruce Lairson (at his own company); Stephen Streiffer (Associate Lab Director, Argonne National Lab); Judith Driscoll (professor at the University of Cambridge); Connie Wang (director in the corporate CTO office at Applied Materials); and Darrell Schlom (professor at Cornell University).

In addition to his vibrant research activity at Stanford, Geballe found time to provide much needed leadership of the materials research community. In the late 1970s, after ARPAs Interdisciplinary Laboratory program had been transferred to NSF as the Materials Research Laboratory (MRL) program, the guidelines for those university research programs began to change. The initial focus on seeding new people and new research and supporting central facilities to benefit all research was giving way to an emphasis on large, focused research projects called Thrusts. The changes required new leadership for CMR, and Geballe stepped up to meet that challenge. He served as director of CMR with distinction for 12 years, from 1976 to 1988.

Geballe has received numerous awards during his long and distinguished career. Among many other accolades, he received the 1970 Oliver E. Buckley Condensed Matter Prize from the American Physical Society "for experiments that challenged theoretical understanding and opened up the technology of high-field superconductors;" was elected to the National Academy of Sciences in 1973; and received the von Hippel award from the Materials Research Society in 1991.

Fritz B. Prinz, 2003. (MSE Dept.)

Friedrich B. Prinz ~ From Metal Physics to Fuel Cells, Catalysts and Capacitors

Friedrich "Fritz" Prinz came to Stanford in 1994 as a full professor in the Department of Mechanical Engineering with a joint appointment in MSE. His appointment was a part of an effort to form stronger links between the MSE department and other Stanford engineering departments engaged in materials-related work. Having worked on the mechanical properties of metals at the University of Vienna and MIT in the 1970s, and having emphasized materials properties and selection in his later work on design and manufacturing at Carnegie Mellon University, he was ideally suited for a joint appointment between Mechanical Engineering and Materials Science.

Prinz built a strong research program in manufacturing at Stanford that focused on rapid prototyping, a system of fabricating prototypes of different sizes and shapes and with different materials for rapid assessment of proposed designs. He developed shape deposition manufacturing, wherein complex parts could be created by computer-controlled sequential deposition and removal of material layers, a precursor to the now popular field of additive manufacturing. This eventually led him to establish the Nanoscale Prototyping Laboratory at Stanford, which he directs. He and his colleagues employ a wide range of nano-fabrication technologies to build prototype fuel cells, solar cells and batteries to test new concepts and novel material structures.

Fritz Prinz was born on March 16, 1950, in Vienna, Austria. His father worked as an agricultural engineer after the Second World War until his untimely death in the late 1960s. His mother worked in the home until she was widowed; she then found a job with Austria's Telecom. In 1968, Prinz graduated from high school in his small town of Gmünd, not far from the border of what is now the Czech Republic. After completing his mandatory military service in the Austrian Army, he enrolled at the University of Vienna to study physics and mathematics. He graduated with a Ph.D. in physics and a minor in mathematics. His research was focused on understanding the behavior of point defects, dislocations and their relation to the mechanical properties of metals. His research was guided by Professors Lintner (experimental physics), Karnthaler (electron microscopy, Kirchner (modeling) and Schöck (theoretical physics).

After serving a few years at the University of Vienna as a university assistant in the Department of Solid State Physics, Prinz joined MIT's Mechanical Engineering Department as a postdoc working with Professor Ali Argon. Prinz and Argon developed the first two-parameter dislocation theory for the mechanical behavior of sub-grain forming metals. He left MIT in 1982 to join Carnegie-Mellon University (CMU) as a junior faculty member in Mechanical Engineering with a courtesy appointment in Materials Science and Engineering. While at CMU, Prinz developed an interest in design and manufacturing, which led to his appointment as director of the CMU Engineering Design Research Center, a National Science Foundation Engineering Research Center. In his classes at CMU, he introduced Lagrange invariance as a means for quantifying forces in manipulators, and used Michael Ashby's materials selection approach as a method for engineering design.

Prinz was attracted to Stanford by the opportunities in micro/nano-scale design, fabrication, and materials engineering. Here he explores nanoscale phenomena for improved energy conversion and storage. As a solid-state physicist by training, Prinz leads a group of doctoral students, postdoctoral scholars, and visiting scholars in addressing fundamental issues on energy conversion and storage at the nanoscale. In his laboratory, a wide range of nano-fabrication technologies are employed to build prototype fuel cells and capacitors with induced topological electronic states. He is testing these concepts and novel material structures through atomic layer deposition, scanning tunneling microscopy, impedance spectroscopy and other technologies. In addition, the Prinz group uses atomic scale modeling to gain insights into the nature of charge separation and recombination processes. Prinz is also interested in learning from nature, particularly understanding the electron transport chain in plant cells. The Prinz group, in collaboration with biologist Arthur Grossman, were the first to extract electrons directly from plant cells subjected to light stimulus.

Prinz's teaching is centered on introducing materials concepts to mechanical engineers and quantum mechanical concepts to materials scientists. His teaching includes courses on "Mechanical Engineering Design," "Additive Manufacturing—From Fundamentals to Applications," and "Fuel Cell Science and Technology." With his former students Ryan

O'Hayre and Suk Won Cha and a former postdoc, Whitney Colella, he wrote *Fuel Cell Fundamentals*, a textbook widely used for courses related to energy conversion.

Prinz's current research focuses on minimizing the usage of noble metal catalysts for the all-important Oxygen Reduction Reaction. This reaction is rate controlling for a range of chemical reactions. In parallel, the Prinz group has been exploring topological electronic states for modulating transport properties in quantum confined material structures.

Prinz now serves as the Leonardo Professor in the School of Engineering with a joint appointment in Mechanical Engineering and Materials Science & Engineering. He is also a Senior Fellow at the Precourt Institute for Energy at Stanford. He was chair of the Mechanical Engineering Department from 2003-2013. He currently serves as the Presidential Appointee of the Visiting Committee for Mechanical Engineering at MIT, and as International Chair for the Faculty of Science and Engineering at the Norwegian University of Science and Technology in Trondheim, Norway. He is a corresponding member of the Austrian Academy of Sciences and a fellow of the American Association for the Advancement of Science. Among other recognitions, Prinz has delivered numerous named lectures around the globe, including the Sir Christopher Hinton Lecture for the Royal Academy of Engineering.

Prinz is married to his wife Gertrud and has two children; his daughter graduated from Berkeley, his son from Stanford.

Andrew Spakowitz ~ Modeling of Soft Materials and Biological Processes

Andrew Spakowitz arrived at Stanford University in the fall of 2006 as an assistant professor in Chemical Engineering. He has developed a research group of approximately five graduate students who utilize analytical theory and computational modeling to address soft materials and biological processes. After being promoted to associate professor in 2014, Spakowitz was appointed as a joint faculty member in Materials Science & Engineering. His research group tackles a range of problems involving polymer self-assembly and molecular motion, in order to elucidate the impact of polymer organization on macroscopic properties and phenomena.

Spakowitz was born in Neenah, Wisconsin, in 1975. At the age of three, he was diagnosed with Ewing Sarcoma when a tumor was found to be growing on the base of his spine. The treatment involved surgery, chemotherapy, and radiation therapy. This treatment left him temporarily unable to walk, but he gradually regained his mobility. His time in primary school was, thankfully, uneventful in comparison to his earlier years. In high school, Spakowitz was a good student and spent considerable time playing piano.

He enrolled as an undergraduate at the University of Wisconsin in 1994. During the UW application process, he auditioned in the Music Department by performing a piano piece, and he had every intention of studying music composition. However, the open audition led him to the conclusion that he would be at best a middling performer in the highly competitive program. He also felt that pursuing a career in music would diminish his enjoyment of music.

Thus, Spakowitz came to UW with no concrete plans for his future studies. Inspiration came from his freshman roommate, who suggested that Spakowitz was not intelligent enough to tackle his own field of study—Chemical Engineering. Thereafter, Spakowitz would enroll in the standard science and math courses necessary to begin the curriculum towards an engineering degree. In addition, he joined the rowing team as a coxswain-- the loud person in the back of the boat who yells commands to the rowers (and wakes up frat boys for 5 am practice on Lake Mendota).

Spakowitz eventually was accepted to upper division work in Chemical Engineering despite

Andrew Spakowitz (A. Spakowitz)

some struggles during his freshman-year studies (particularly in introductory calculus). Upon transitioning to the core engineering classes, he hit his stride and began to excel as a student, particularly in his studies of thermodynamics. Professor Juan de Pablo took note of Spakowitz within his multi-component thermodynamics course, which at University of Wisconsin had more than 100 students enrolled. This led to Spakowitz working in de Pablo's lab as an undergraduate researcher and as an editor of the Juan de Pablo and Jay Schieber textbook *Molecular Thermodynamics*. During his undergraduate studies, Spakowitz spent two semesters as a co-op student at a polyethylene plant in Clinton, Iowa.

His co-op work experience was extremely valuable, but he eventually opted for graduate school in Chemical Engineering at Caltech. He met his future wife, Sarah Heilshorn, during the recruiting visit, as she was a first year graduate student at Caltech participating in graduate recruitment. Later, Heilshorn would be the first woman to join the MSE faculty at Stanford. The University of Wisconsin and Caltech could not have been more different from each other in terms of size, intellectual focus, and interests of the students. Spakowitz found the smaller size and fundamental focus of Caltech was ideal for his development during his graduate career. He joined Professor Zhen-Gang Wang's Caltech research group to study the physical behavior of soft polymeric materials. His specific project was developing mathematical models to predict the behavior of semiflexible polymers, which include many synthetic and biological polymers.

During his graduate career, Spakowitz found exact mathematical solutions for the statistical behavior of the standard model for semiflexible polymers, namely the wormlike chain model (WLC). This model has existed for more than 60 years, and its mathematical solutions can be applied to a range of biophysical and soft-material problems, and represent a major advancement in understanding polymer behavior. Utilizing WLC has enabled Spakowitz to tackle a number of problems throughout his career.

Spakowitz was hired as an assistant professor at Stanford in 2004, but he delayed his start date until fall of 2006 in order to be a postdoctoral scholar with Professors Carlos Bustamante and George Oster at U.C. Berkeley. That experience involved extensive interactions with experimental biophysicists in the Bustamante lab. This transition to applying theory to explain biological phenomena, particularly in living systems, was crucial to Spakowitz's development. This theme of directly engaging *in vivo* biological processes is central to Spakowitz's present-day research program, and the nature of his work has been profoundly influenced by these experiences.

Spakowitz married Sarah Heilshorn in June of 2005 in Grand Rapids, Ohio, near where Heilshorn grew up. They live in Mountain View with their two sons, Cyrus and Oscar, who were born in 2011 and 2012.

Chapter 18

Consulting Professors: Enriching the Department's Teaching and Research Programs

For nearly 50 years, the department has benefitted from the contributions of individuals holding the title of consulting professor. They serve an important role, helping to guide graduate students and post-doctoral students in their research. In some cases, they are actively involved in the teaching program. Most commonly, consulting professors hold full-time positions in local companies or laboratories, or are recently retired from such organizations, and are typically not paid for their contributions to the department. Usually closely associated with one or more faculty members in the department, they contribute to the teaching and research programs of that particular group. In some cases, as described below, consulting professors also have made major contributions to the development of the department by introducing completely new facilities and capabilities for advanced materials research.

Below is a list of the individuals who have served for at least three years as consulting professors in the department.

Farid Abraham, 1970-1981

Robert I Jaffee, 1975-1990

Helmut R. Poppa, 1975-1986

John Stringer, 1978-2000

Jeffrey Wadsworth, 1981-2004

Arden Sher, 1982-2000

Paul A. Flinn, 1986-2007

Timur Halicioglu, 1986-2007

David Redfield, 1986-1998

Henry Wise, 1986-1992

Michael A. Kelly, 1991-

Thomas Marieb, 1997-2001

Jamshed R. Patel, 1998-2007

Baylor Triplet, 2001-

Robert M. White, 2004-

Turgut Gur, 2005-

Robert E. Fontana, 2007-2014

Wendelin J. Wright, 2007-2011

Rene Meyer, 2007-2011

Geraud Dubois, 2010-

Rommel Noufi, 2010-2016

Kristin Persson, 2014-

Khalil Amine, 2018-

All of the consulting professors listed above have enriched the department in some way, either through teaching or research. Three people stand out as having made especially important contributions: Jeffrey Wadsworth, Paul Flinn, and Michael Kelly.

Jeffrey Wadsworth ~ Superplasticity, Brilliant Teaching and International Leadership

Jeffrey Wadsworth came to the department in 1976 as a post-doctoral student to work on the superplastic properties of ultra-high carbon steels with Oleg Sherby. He had received a B.S. degree in 1972 and a Ph.D. in 1975 in metallurgy from Sheffield University. Sheffield awarded him a Doctor of Metallurgy degree in 1991 for his published work, and an honorary Doctor of Engineering degree—the highest recognition conferred by that university—in 2004.

During Wadsworth's 40-year collaboration with Sherby, they wrote nearly 90 papers together, mostly on the superplastic properties of ultra-high carbon steels, including their most highly cited paper, "Superplasticity—Recent Advances and Future Directions," published in *Progress in Materials Science* in 1989. In addition, together with T.G. Nieh, another graduate of the MSE department, they published their well-regarded book, *Superplasticity in Metals and Ceramics*, in 1996.

Jeffrey Wadsworth (J. Wadsworth)

By 1980, Wadsworth had moved to the Lockheed Missiles and Space Company in Palo Alto and was working his way up in that organization. He continued to maintain his contacts with Sherby and the department through his appointment in the department, first as a consulting assistant professor in 1981 and later a consulting professor. By the mid-1980s, Wadsworth was taking on part-time teaching assignments in the department in addition to his continued collaboration with Sherby. In particular, he taught the undergraduate engineering course, E 51, "Materials Technology for Structural Applications," where he brought to the classroom real-world applications from his work at Lockheed. Wadsworth's unique workplace perspective, added to his natural gifts as a speaker, resulted in his being judged the top teacher in a department that prided itself with having excellent teachers among its faculty. For years, Wadsworth was ranked the top teacher in the department based on the student reviews. He later took on the teaching of MSE 205, "Strength and Microstructure," where he continued to be the highest rated teacher in the department.

Wadsworth's ability to articulate important issues in materials science and technology led to his rapid rise in Lockheed and to a national reputation. In 1992, he joined the staff of the Lawrence Livermore National Laboratory and became Deputy Director of Science and Technology for the laboratory in 1995. He served in that capacity until 2002, when he joined the Battelle Memorial Institute in Columbus, Ohio. Soon after joining Batelle, he served as a member of the White House Transition Planning Office for the U.S. Department of Homeland Security. The following year he was named Director of Oak Ridge National Laboratory, the U.S. Department of Energy's largest multipurpose science laboratory and one of the national laboratories managed by Battelle. Wadsworth served in that role until 2007, when he moved to the

Battelle headquarters in Columbus and led Battelle's Global Laboratory Operations business, overseeing the management of six national laboratories of the U.S. Department of Energy. Finally, in 2009, he was appointed President and Chief Executive Office of Battelle Memorial Institute, a position he held until his retirement in 2017.

For his national leadership in materials science and technology, Wadsworth has received a number of prestigious awards. These include election to the rank of Fellow of three technical societies; election to the U.S. National Academy of Engineering in 2005 and to the Chinese Academy of Engineering in 2011; six honorary doctorates; two honorary professorships in China; the 2013 Acta Materialia Award in Materials and Society; and an Honorary Membership in ASM International in 2017.

Paul A. Flinn (P. Flinn)

Paul A. Flinn ~ A Creative Force in Mechanical Properties of Thin Films

Paul Flinn was a senior Staff scientist at Intel Corporation in Santa Clara, California, when he joined the department as a consulting professor in 1985. He had been at Intel since 1978 and was engaged in work on the reliability of microelectronic devices, but the need for him to focus on short-term, company-specific problems did not give him the freedom to pursue fundamental scientific questions, as he had done before as a professor at Carnegie Mellon University. Seeing the need for fundamental research on microelectronic reliability, Intel generously allowed Flinn to spend time at Stanford, providing funding to support his work. Flinn collaborated primarily with John Bravman, who was also starting to work on mechanical properties of thin films and reliability of microelectronic devices, but Flinn largely operated as an independent researcher at Stanford with research assistants of his own, students who were formally advised by Bravman.

At the time, research on the mechanical properties of thin films was just beginning and there were few tools available for such work. In particular, although the Stoney relation (showing how the stress in a film on a substrate could be determined from the change in curvature induced by that stress) had been known for a long time, instruments for making such curvature measurements were not widely available. Using his strong background in experimental physics, Flinn set out to build an instrument for monitoring the substrate curvature and the corresponding biaxial stress in a film as a function of temperature or other processing conditions. With the new curvature instrument, Flinn, with Donald Gardner, a student in electrical engineering, and the author of this history, published a paper in 1987 on the measurement and interpretation of stress in aluminum-based metallization as a function of thermal history. That paper was the first that Flinn had published since becoming a consulting professor, and it became his second most highly cited paper. Many new wafer curvature instruments followed from that original work, leading to a number of publications and dissertations in the department on the mechanical properties of thin films.

Flinn's pioneering work also led to similar efforts by researchers at the Max Planck Institute in Stuttgart, Germany, with whom Flinn was also collaborating at that time.

While the substrate curvature technique provided a powerful way to study the stresses in blanket films of uniform thickness, it was essentially powerless to determine stresses in buried interconnect lines or other patterned structures. For that, Flinn, with his Intel colleague, Chiang Chien, developed a clever technique for using x-ray diffraction to measure the stresses in passivated interconnect lines. Use of that technique provided a much fuller picture of both the stresses that develop in passivated interconnect lines and how those stresses lead to failure. The method that Flinn and Chien developed was used by many students in the department engaged in work on the reliability of interconnect structures.

One set of problems that was plaguing the microelectronics industry at the time was the voiding and failure of metallic interconnect lines caused either by stresses in the lines or by electromigration conditions. In an effort to directly observe the voids responsible for these failure processes, Flinn, with Jonathan Doan, a graduate student formally under John Bravman, designed and built a high voltage scanning electron microscope (HVSEM) with a large specimen chamber that permitted electromigration tests to be conducted on realistic test structures while simultaneously allowing observation of the passivated (buried) interconnect lines at high magnification. They did this by modifying a scanning transmission electron microscope operating with high-energy electrons (120keV), which allowed for imaging of the metal lines and the voids within them through detection of the backscattered electrons. With this technique, Flinn and Doan were able to directly observe the voids created in metal lines by electromigration processes while measuring the corresponding changes in electrical resistance at high temperatures. The use of this new instrument led to a whole series of publications by Flinn, Doan and Bravman that greatly advanced the understanding of electromigration failure processes in interconnects in integrated circuits. Jonathan Doan later served as an Acting Assistant Professor in the department, as reported elsewhere.

Paul Flinn was born in New York City on March 25, 1926. After graduating from Gorton High School in Yonkers, New York, in 1943, and serving in the U.S. Navy for two years, he attended Columbia College in New York, graduating with a bachelor's degree in 1948 and a master's degree in 1949, both in physics. He moved on to MIT, where he started his career in materials science. He studied with B.L Averbach and received his Sc.D. degree in metallurgy in 1952 for his work on the local atomic arrangements in metallic solid solutions using diffuse x-ray scattering.

After serving on the physics faculty at Wayne State University for a year, Flinn joined the Westinghouse Research Laboratory in Pittsburgh, Pennsylvania, where he worked for nearly a decade. It was during that time that he published his most widely cited paper that showed how screw dislocations in certain intermetallic alloys could spontaneously move into locked configurations and cause the strength of that alloy to increase with temperature, rather than decrease with temperature as all other high temperature alloys do. His discovery provided a fundamental explanation of why nickel-based superalloys retain and even increase their strength at elevated temperatures. By the mid-1960s, Flinn had joined the faculty in physics at Carnegie Tech (later, Carnegie-Mellon University) where he stayed until his move to Intel in 1978.

Flinn has been recognized for his work through his election as a Fellow of the American Physical Society and through his listing in *Who's Who in Science and Engineering in America* and in *Who's Who in the World*. Since his retirement in 2007, he has lived in Hawaii, where he continues to work on the structure of glasses using molecular dynamics simulations.

Michael A. Kelly ~ X-ray Photoelectron Spectroscopy and Generous Guidance of Students

Michael Kelly was responsible for bringing the technique of x-ray photoelectron spectroscopy (XPS), sometimes called Electron Spectroscopy for Chemical Analysis (ESCA), to Stanford. XPS is a surface-sensitive quantitative spectrographic technique for measuring the elemental composition and the chemical and electronic states of elements residing at the surfaces of materials. XPS spectra are obtained by irradiating a material with a beam of x-rays while simultaneously measuring the kinetic energy and number of electrons that escape from the surface. Kelly was one of the engineers at Hewlett-Packard Laboratories who had developed the first commercial monochromatic XPS instrument in 1969. His work at HP was done in collaboration with Kai Siegbahn of Uppsala University in Sweden, who had invented the technique in the 1950s. Siegbahn would receive the Nobel Prize for Physics in 1981 for his extensive efforts to develop XPS into a useful analytical tool. Stig Hagström, a professor in the MSE department in the late 1980s and 1990s and chair of the department during that time, had been a graduate student with Siegbahn in the early 1960s. It was Kelly's connection to Hagström that brought Kelly to Stanford in the mid-1980s.

Near the end of his eight years at HP Labs, Kelly, with three others, formed Surface Science Laboratories in Mountain View in 1974 to design and build XPS spectrometers and provide surface analytical services to the semiconductor industry. The company got its start by buying XPS equipment from HP Labs. The company prospered in the 1970s and eventually employed nearly 100 people. By the early 1980s Surface Science Labs merged with Kevex Corporation, a publicly traded company engaged in the design and marketing of electron spectroscopy instrumentation for chemical analysis. Kelly became Chief Operating Officer and later president of the combined company, but in 1984 the combined company was sold to Vacuum Generators (VG), the UK transmission electron microscope manufacturer. At this time, Kelly began to work with Hagström at Stanford.

When VG later failed to prosper in the surface analysis business, Kelly bought the analytical lab back from the unsuccessful business with the hope of re-starting a surface science company. When that effort did not succeed, Kelly turned more of his attention to developing an XPS laboratory at Stanford. At the urging of Hagström and Robert Huggins, Kelly essentially donated his XPS equipment to Stanford and volunteered to set it up and establish an XPS characterization facility at the university. It would later become one of the key facilities within the Stanford Nanocharacterization Laboratory in the Geballe Laboratory for Advanced Materials. By 1991, Kelly's work at Stanford was formally recognized by his appointment as a consulting professor.

Kelly would prove to be a great resource for faculty, students, and post-docs engaged in thin film research. With these researchers he published dozens of papers on the XPS technique itself and on the use of XPS to characterize the growth of thin films of all kinds, especially diamond and diamond-like thin films, which were attracting much attention at the time. His most highly cited paper involved an effort to characterize diamondoid monolayers with XPS. Co-authored with Z.-X. Shen of the Applied Physics Department and Nicholas Melosh of the MSE Department, among others, it was published in *Science* in 2007. In addition to his research contributions and his leadership in establishing an XPS laboratory at Stanford, Kelly was actively involved in teaching. Soon after his appointment as a consulting professor, he took over the teaching of MSE 312, "New Methods for Thin Film Synthesis," a course that had been initiated by Hagström. He taught that course with distinction for several years.

Michael A. Kelly, 1988.
(Stanford News Service)

Michael Kelly was born in Roswell, New Mexico on December 14, 1936. He grew up in Roswell and graduated from high school there in 1955. With an ROTC scholarship from the Navy, he was able to enroll at UCLA, where he studied electrical engineering and graduated with a B.S. degree in 1959. He then spent three years in the Navy working at the Brooklyn Navy Yard. While there, he attended Brooklyn Polytechnic Institute, graduating in 1963 with an M.S. degree in electrical engineering. By then he was in the doctoral program in the physics department at the University of California at Berkeley. He received his Ph.D. in nuclear physics from Berkeley in 1968 and immediately joined HP Laboratories in Palo Alto. There he ran HP's R&D and marketing efforts, which led to the first commercial XPS spectrometers in 1972.

Kelly continues to serve as a consulting professor in the department, where he provides sage advice and guidance on XPS and other characterization methods to students, post-docs and faculty from across the advanced materials community. He is perhaps the only person from the department to have been honored with a tribute published in the Congressional Record. Representative Anna G. Eshoo introduced a tribute to Michael Kelly on September 24, 2008, on the occasion of his "retirement" from the department.

Chapter 19

Faculty Appointments of the 1980s

By the mid-1980s, the department had gone nearly a decade without making any faculty appointments. This dearth of appointments was caused by both the almost non-existent undergraduate enrollment of the 1970s and by the perception of leaders in the School of Engineering that the department had been poorly led and had failed to fully embrace important materials issues in silicon technology. A faculty search was approved at that time with the aim of appointing a faculty member who would work on problems related to silicon technology and who would form strong connections with faculty in Electrical Engineering. That effort led to the appointment of John Bravman, who had received all of his degrees from Stanford's MSE department, but had actively collaborated with James Plummer, Krishna Saraswat, and others in Electrical Engineering during his doctoral studies.

In the latter part of the 1980s, the issue of department leadership was addressed by an external search for a new department chair. That led to the appointment of Stig B. Hagström as professor and department chair. Hagström had been at the Xerox Palo Alto Research Center, where he had strong connections with the faculty in Electrical Engineering at Stanford. Some of the turmoil in the department had subsided by this time, so the department was permitted to conduct a search for a faculty member to fill the gap in the area of kinetics that had been created in 1980 when Marshall Pound took medical retirement. As many of the senior faculty members were beginning to retire, the department also initiated a practice of appointing acting assistant professors to help maintain and enrich the program.

John C. Bravman ~ Electronic Materials Science and University Leadership

John Bravman joined the MSE faculty in January of 1985, after having received all of his degrees from the MSE department. He has been the only MSE student to ascend to a faculty position in MSE for the past 50 years. At the time of his appointment, the department had been struggling to find a way to fully embrace silicon technology. Having worked jointly with Robert Sinclair in the MSE department and James Plummer in the Electrical Engineering department on critical issues in integrated circuit technology, Bravman was ideally suited to help the department move into materials research areas pertinent to silicon technology. He went on to establish a vibrant research program on electron microscopy of interfaces in electronic materials; mechanical behavior of microelectronic materials; electromigration failure processes; and high temperature superconducting materials, before taking on major leadership positions in the department, school, and university. Along the way he also became the department's much-recognized outstanding teacher. After having served in some of the highest levels of leadership at Stanford, Bravman accepted appointment as president of Bucknell University in 2010.

Bravman's 35 years at Stanford began with an accident of friendship and what he described as "a bitter rejection." He was always interested in science, and so, as a New York native, he did the expected and applied to MIT, along with a handful of other eastern schools. In the end, however, MIT did not see in him what they thought they needed in their student body, and he was rejected. As he has said, "if you were a 'techie,' and especially if you grew up in the east, this is where you wanted to go to college." Fortunately, Bravman also applied to a school that was almost unheard of in his sphere – Stanford – but only because a good friend from high school had vacationed in California the summer after their junior year.

John Bravman at the TEM, 1986. (J. Bravman)

Because Stanford's financial aid package was generous enough to allow Bravman to attend, he arrived on campus on September 18, 1975. After September 1978, he was never off the Stanford campus for more than 15 consecutive days until July 2010 when he left to become Bucknell's president.

Bravman's academic start at Stanford was not auspicious, as he has recounted to countless students over the decades. Thankfully, everything changed in the spring of his sophomore year when, almost by accident, he took Engineering 50, "Introductory Science of Materials," taught by a new, young professor, Robert Sinclair. Bravman was fascinated by the subject, admired Sinclair deeply, changed his major from electrical engineering to materials science, and asked Sinclair to be his advisor. The following fall, he enrolled in Sinclair's course on structures and crystallography, then labelled MS&E 180, and which included both graduate and undergraduate students. Not too many students enjoy crystallography, but Bravman came to love it, and later taught it for many years.

The power of mentorship came to the fore, and Sinclair convinced Bravman to try for the Ph.D. program, since "he would be paid, not pay, and because it meant he could spend five more years at Stanford." The summer after Bravman's senior year, Sinclair also arranged a job for him at Fairchild Semiconductor, working in their materials analysis lab for Sinclair's friend, Stefan Justi. Bravman worked fulltime that summer, then returned to school in the fall to begin his doctoral studies while continuing to work part time at Fairchild—perhaps a portent of his continuing work with companies in the valley for decades to come.

Because his dissertation project involved silicon process technology, not a departmental strength, Bravman started building relationships with faculty in Electrical Engineering, especially Jim Plummer, Krishna Saraswat, Simon Wong, Jim Meindl, and Jim Gibbons. This network of collaborators turned out to be important over the entire course of his career. Somewhat later, he also worked for a period in high temperature superconductors, following the breakthrough developments in that field, and thus came to work with Ted Geballe and Mac Beasley in Applied Physics.

While a graduate student, Bravman began to be interested in public speaking. He faced his first conference presentation at the 1982 Fall MRS meeting, then still in the Boston Park Plaza. Despite having enormous stage fright, or perhaps because of it, he was dedicated to giving a good talk. It went well –as this author can attest—and upon returning to campus, Bravman started giving guest lectures, especially in E 50, the course that first drew him to the department.

In Bravman's final year as a doctoral student he had several offers from the likes of Fairchild and IBM, but the MSE department decided that it had to build a faculty presence squarely in silicon technology. It secured two billets for this, one of them coming as a "bulge" billet based on Artie Bienenstock's role at SSRL. Bravman recalls that Jim Plummer, by then a clear superstar in EE and one of his dissertation advisors (along with Professors Sinclair and Saraswat) told him that he wanted Bravman to apply for one of those positions. Bravman remembers being incredulous at this suggestion.

Bravman started his professorial career on the faculty in the Fall of 1984 as an acting assistant professor, teaching what was then MSE 202A, a laboratory course focusing on metallography and x-ray analyses. Cutler Shepard was sometimes around, and he told Bravman how he used to teach a similar course. Bravman's dissertation was not finished yet, so during this first term of teaching he was also writing furiously and preparing for his defense. He presented his dissertation talk on October 24 and turned in his thesis a few minutes before 5pm on December 7, Pearl Harbor Day, which was also the last day that term for dissertations to be accepted.

Most dissertations then were still prepared by typists, but Bravman's was one of the first to be prepared on a computer and printed using a laser printer, at the time a rare device, even at Stanford. He used a typographic system named "Scribe," running on a large computer, but admits he should have used the early version of Don Knuth's now-ubiquitous TeX software. (Fast forward 34 years: in late 2018, Knuth was featured in a major piece in the *New York Times* and was aptly referred to as "the Yoda of Silicon Valley.")

Over time, Bravman developed a research program based on silicon materials processing, thin film structures, and the like; his first doctoral student was David Paine, who is still on the faculty at Brown University. Later, Bravman also worked on high temperature superconductors. Bravman built a lengthy collaboration with the author of this piece around his and David Barnett's growing interest in the mechanical behavior of thin films. Paul Flinn, an adjunct faculty member who was an Intel engineer, was an important part of this work. Idiosyncratic in the extreme, but highly capable and experienced, Flinn was crucial to the collaboration, as they designed and built several pieces of specialized equipment for this and related projects. Equipment built included a standard wafer curvature machine (which was still fairly rare); the world's only high temperature wafer curvature machine capable of operating at well over 1000C; a rotating anode x-ray system for thin film analyses; a micro-tensile testing system in which lithographic procedures were employed to make micron-scaled "dog-bones" for tension testing; a cantilever microbeam deflection system using a nano-indentation system (itself a concept championed by a former graduate student, Warren Oliver); and a high voltage scanning electron microscope (accomplished by cutting a TEM in half and changing the optics) that could accommodate samples in a test rig for in-situ electromigration of passivated aluminum and copper.

Bravman's work in this area, as well as his work in high temperature superconductors, earned him tenure in 1991. He had also become an effective teacher and mentor, earning the School of Engineering Distinguished Advisor Award in 1987; the Excellence in Teaching Award from the Society of Black Scientists & Engineers (which Robert Sinclair had won earlier); and in 1989 the Walter J. Gores Award for Excellence in Teaching, Stanford's highest award for teaching. He also won the 1991 ASM national Bradley Stoughton Award for Teaching. He was presented this latter award at a black-tie ceremony, where Bravman also happened to meet David Packard; many years later, Bravman met Packard again at the dedication of Gates Computer Science building.

Early on, Bravman started getting involved in "administration," as it typically is called, and on his first day with tenure—September 1, 1991—he also became associate chairman of the department, serving under Stig Hagström, the former Xerox PARC researcher who had been hired to be chairman after Richard Bube. Over time, Bravman would be asked to take on a growing number of positions at the department, School of Engineering, and University level, sometimes holding two such positions at the same time.

Bravman was promoted to full professor after four years, and soon thereafter became the Bing Centennial Professor. He worked extensively with Stanford alumnus Peter S. Bing, himself a legend on campus for his decades of service on the Board of Trustees and for his campus-wide philanthropy. Bravman always thought of the professorship as an incredible honor for him and his family.

Bravman moved onto campus in 1999, at 225 Santa Teresa Lane, and started hosting a yearly Christmas dinner party that was deeply enjoyed by all. He always asked Dave Barnett and the author of this retrospective to share some stories from "the old days," which he thought the younger faculty would enjoy. Some years Bravman had to remove all the furniture from his living room so that a group numbering more than 30 faculty, students, and staff, plus spouses, could squeeze into his home.

In 2005, John Bravman married Wendelin J. Wright, who, like Bravman, had received all of her degrees from the MSE department. Wright was serving the MSE department as an acting assistant professor around that time. She is currently the Heinemann Family Professor in the departments of Mechanical Engineering and Chemical Engineering at Bucknell University. They have two sons, Cole and Cooper.

John Bravman, then Vice Provost for Undergraduate Education and Dean of the Freshman/Sophomore College, speaking at Freshman Convocation, in Stanford's historic Quadrangle, ca. 2005. (Stanford News Service)

John Bravman and Wendelin Wright, at a Stanford commencement dinner, 2008. (J. Bravman)

The pinnacle of Bravman's administrative work was his eight years' service as the Senior Associate Dean in the School of Engineering, overlapping with eleven years as Stanford's Freeman Thornton Vice Provost for Undergraduate Education. In Engineering, he served under deans Jim Gibbons; John Hennessy; and his former doctoral co-advisor, Jim Plummer. The Terman Engineering Building, built during Bravman's undergraduate days, had to be torn down in 2011 owing to a flaw in the laminated beams from which it was constructed, but Bravman was able to have the nearby grove dedicated in honor of Gibbons. To this day, Gibbons Grove still holds a stone tablet inscribed with Bravman's tribute.

As Vice Provost, Bravman served under Gerhard Casper and then John Hennessy, two very different but truly remarkable men who were Stanford's presidents from 1992-2000 and 2000-2016, respectively. Casper had brought a renewed focus on undergraduate education, but his unexpected resignation in 2000 left Hennessy, who had been Casper's provost for just a short time, and who went on to become president, the task of designing and completing the Campaign for Undergraduate Education, the university's first campaign of one billion dollars focused solely on undergraduate education. As part of the campaign, Stanford ran the "Think Again" road show through 12 cities in just one year. Bravman was active in the subsequent "Leading Matters" campaign, but it was "Think Again" that really capped off his Stanford career and prepared him for a university presidency. When Bravman left Stanford for Bucknell, he was awarded the university's highest award for service, the Kenneth M. Cuthbertson Award for Contributions to Stanford University. Only two people, as of this writing, have won both the Gores and the Cuthbertson. This was great honor for Bravman, but he was most pleased that it brought distinction and notice to the MSE department.

As Bravman left Stanford, his 35-year career there was recognized with his appointments as the Bing Centennial Professor, Emeritus, and the Freeman-Thornton Vice Provost for Undergraduate Education, Emeritus. These titles signify the collaborations and hard work of which he is most proud.

Stig B. Hagström ~ New Department Leadership and Renewed Industry Support

By the mid-1980s, the department was aging, and many faculty members were nearing retirement. Richard Bube and this author had been chair and associate chair, respectively, for a decade. In addition, the reputation of the department within the school had declined, in part, because the department had not yet fully embraced materials science problems in silicon technology. So, when James Gibbons became the Dean of Engineering in 1984, he started discussions that led to a search for a new department chair from outside of the department and the university. That search, begun in 1985 and concluded in 1986, resulted in the appointment of Stig B. Hagström as professor and chair of the MSE Department, effective September 1987.

Hagström, who had been Manager of the General Sciences Laboratory at the Xerox Palo Alto Research Center, had already been closely associated with Stanford since 1976 as a consulting professor of electrical engineering. He had worked with Ingolf Lindau and William Spicer of the Electrical Engineering Department to establish one of the first beam lines at the Stanford Synchrotron Radiation Project (SSRP) in 1974.

During the five years Hagström was on the faculty, he introduced a new course entitled "New Methods in Thin Film Synthesis" that he later co-taught with Donald L Smith, a former colleague at Xerox PARC who was then writing a textbook on the subject. With others, he introduced a seminar course on "Thin Film Science and Technology," and even taught one of the core courses in the department, MSE 205, aves and Diffraction." The word "even" is used here because Hagström's primary focus in the department was elsewhere. He was concentrating on resurrecting the Industrial Affiliates program, which had fallen by the wayside, and on improving the image of the department within the school and university. He also introduced surface science to the department, mainly through his close collaboration with Michael Kelly, who brought Electron Spectroscopy for Chemical Analysis (ESCA), to the department and Stanford.

As a graduate student at Uppsala University in Sweden, Hagström had worked on ESCA with Kai Siegbahn, the inventor of the technique. Hagström maintained his connection to Sweden throughout his time at Stanford, and later returned to Sweden to help with the development of higher education in Sweden, based on the Stanford model. Still later, he returned to Stanford and helped to establish the Wallenberg Research Link (WRL) and the Wallenberg Global Learning Network (WGLN) to foster collaborations between Stanford and Swedish universities. Stig B. Hagström was born in Esperyd, Sweden on September 21, 1932. He received his B.S. degree in 1958 and his Ph.D. in 1964, both from Uppsula University. As a graduate student he was instrumental in helping to develop x-ray photoelectron spectroscopy (XPS) which was later called Electron Spectroscopy for Chemical Analysis (ESCA). He studied with Kai Siegbahn who would win the 1981 Nobel Prize in Physics for that work. Hagström was one of those who discovered the chemical shift in XPS and established it as a basis for materials analysis.

Hagström did post-doctoral work at MIT before moving to the Lawrence Berkeley Laboratory to work with David Shirley. There he helped to establish the first experimental XPS laboratory in the U.S. His most highly cited paper, "Chemical effects on core-electron

Stig B. Hagström, 2001. (Stanford News Service))

binding energies in Iodine and Europium," with C.S. Hadley, M.P. Klein and D.A. Shirley, came from that work. In 1966, Hagström returned to Sweden to accept an associate professorship at Chalmers University of Technology. It was there that he suggested to William Spicer of Stanford that the high-energy physics storage ring (SPEAR) to be constructed at SLAC should be used as a source of synchrotron radiation for research. That suggestion led to the development of the Stanford Synchrotron Radiation Project (SSRP).

In 1969, Hagström accepted an appointment as professor and chairman of the Science Department at Sweden's newly founded Linköping University. There he started programs at both the graduate and undergraduate levels while still pursuing his own research in the fields of surface physics, surface analysis, and surface coatings with physical vapor deposition techniques. The groundwork he laid in materials science at Linköping has led to its being one of the strongest research environments in Sweden in that field. As Linköping's pro-vice-chancellor from 1970 to 1976, he greatly increased the contacts between the university and local industry, acquiring experience that he would later use at Stanford.

Hagström returned to California in 1976 to lead the Xerox Palo Alto Research Center's General Sciences Laboratory. The lab was already internationally recognized for its work on synchrotron radiation, surface and bulk properties of metals and semiconductors, and many other accomplishments, and would grow in stature under Hagström's leadership. It was then that he renewed his collaborations with Lindau and Spicer, through his appointment as a consulting professor in electrical engineering. Hagström was well known to James Gibbons, Dean of Engineering, when Gibbons sought new leadership for the MSE department a decade later.

During his time in the MSE department, Hagström's major contributions involved the resurrection and re-building of the MSE Industrial Affiliates program, which had fallen into a moribund state. The years of neglect are illustrated by an earlier department chair having lost one of the Industrial Affiliates checks—only to find it months later under a pile of papers. Hagström's upbeat personality and his previous experience in connecting Linköping University to its local industry made him well suited to re-build the Industrial Affiliates program.

Through his international contacts and tireless traveling around the world, Hagström was able to attract nearly 20 companies to become Industrial Affiliates. At that time, a company was asked to contribute $10,000 a year to the department in exchange for early access to the department's research activities and, especially, for early contact with students and post-docs who might be prospective employees. The model had worked well in the 1960s when major companies spent corporate money to maintain contacts with universities but by the 1990s this model had become harder to sustain, as the funding of research in companies became more decentralized. Since the 1990s, companies have been more interested in providing even greater levels of support than in previous decades, but primarily for projects of direct interest to the companies' goals, rather than for general support for the department.

It was during Hagström's time that the MSE department was in the midst of a transformation from a traditional department with emphasis on bulk structural materials to a department with a focus of thin films and nanotechnology. Hagström played an important role in that transformation through his work on surface science. He attracted leaders of that field to the university and introduced Electron Spectroscopy for Chemical Analysis (ESCA) to the Stanford facilities. In addition, with Michael Kelly, who had been appointed as a consulting professor, Hagström established a research group of graduate students that focused on the growth of diamond thin films using plasma-assisted chemical vapor deposition techniques, among other subjects. In the 1990s they published a number of papers on these subjects.

By the early 1990s, Hagström was thinking about contributions he could make in the development of higher education in Sweden. He resigned as department chair in 1991 and became a Stanford Professor (Emeritus), a rank that would allow him to return to Sweden.

Over the course of his career, Hagström had become a friend and confidant to Sweden's royalty and national leaders, so it was not unexpected that he was selected to be the chancellor of the country's entire university system when he returned to Sweden. He served in that capacity from 1992 to 1998, during which time he supervised the restructuring of the country's 37 colleges and universities and assumed responsibility for the education of 150,000 students. He chaired the board of the Swedish National Agency for Higher Education, tasked with developing and introducing a system for evaluating the quality of education and research at Swedish universities and colleges. He also headed a 2000 parliamentary investigation of the Swedish research-funding system. In these assignments, he surely drew on his experience at Stanford and in Silicon Valley.

Having been elected as a member of the Royal Swedish Academy of Sciences in 1992, in subsequent years Hagström was active in the selection of Nobel laureates. He was elected to the Royal Swedish Academy of Engineering Sciences in 1983, and was its chairman from 1993-1996. Even after he returned to Stanford in 1999, Hagström traveled each year to Sweden to attend the Nobel award ceremony.

Hagström returned to Stanford in 1999 to help establish the Wallenberg Research Link (WRL) and the Wallenberg Global Learning Network (WGLN). The WGLN nurtures collaborations between Stanford and Swedish university faculty on using innovative information-technology methods to facilitate the learning process. In 2001, he became co-director of the Stanford Center for Innovations in Learning, an interdisciplinary center that conducts scholarly research to advance the science, technology, and practice of learning and teaching. In that role, he was a key figure behind the creation of Wallenberg Hall, a "technologically agile" home for research in university-level classroom learning through experimentation in new methods of education.

Hagström received many honors and awards for his leadership in science and education, including his election as a member (and later honorary member) of the Royal Swedish Academy of Engineering Sciences and his membership in the Royal Swedish Academy of Sciences; the Royal Norwegian Society of Sciences and Letters; and the Royal Physiographic Society in Lund. He was also recognized with honorary doctorates of engineering by Linköping University and Blekinge Institute of Technology. In addition, he received the Royal Order of the Seraphim, a Swedish Royal Order of Chivalry created in 1748.

Hagström retired to his home in Menlo Park where he lived with his wife, Brita-Stina, before she died in 2006. Together they always warmly received Swedish scientists, politicians, industrialists, and journalists to their home. He died of a stroke on May 28, 2011.

Bruce Clemens ~ From College Drop-out to Top Level Materials Scientist

Standing in a sleet-soaked parking lot in Warren, Michigan, Bruce Clemens was wondering why Robert Sinclair, the eminent electron microscopist that Clemens had invited to give the colloquium at General Motors Research, was telling him about a faculty search in the Department of Materials Science & Engineering at Stanford. "It's in the area of kinetics," Sinclair announced in his proper British accent. As it turned out, that conversation was the beginning of a process that would change Clemens's life.

Clemens had taken a somewhat unusual path to his meeting with Sinclair in the GM parking lot. Born in Houston in 1953, he had been raised in Ohio and California, graduating from Palo Verdes High School. He had gotten the bug for science and engineering early—his father and grandfather were engineers and his mother a chemist. Clemens was unusual for his generation, in that all four grandparents were college graduates; he also was raised to expect that the education and career aspirations of his three very smart sisters were at least, if not more, important than his own. But he had not exactly covered himself in glory in his first venture in higher education at UC Santa Barbara: as an undergraduate, he had dropped out after receiving a failing grade in physics. He had not opened his book all quarter, an approach that had served him well in high school, but which was clearly not appropriate for college level work.

Bruce Clemens (MSE Dept.)

With his tail between his legs, Clemens trudged off to Denver, Colorado, finding work in a machine shop as a tube bender. Here, in the massive deformations involved in bending metal tubes, he had his first exposure to materials behavior. He soon realized that he had not given college his best effort, and began taking night classes at the Denver campus of the University of Colorado, which at the time was in an old bus terminal downtown. (Clemens and his fellow students jokingly called it UCLA: University of Colorado between Lawrence and Arapaho Streets.) He found he enjoyed his calculus, physics and chemistry courses, but was frustrated that they skipped the hardest sections of the books. He transferred to Colorado School of Mines, which the engineers at the machine shop seemed to hold in high esteem. He chose to major in Engineering Physics, since it was physics that had been his downfall and had left him with what he felt was unfinished business.

He continued working in the machine shop until he was offered a position in a research lab studying heterogeneous nucleation of water droplets on coal dust particles under Professor James Brown. Here he had his first exposure to classical nucleation theory. He minored in metallurgy, and he took physical metallurgy from Professor Glen Edwards, a Stanford MSE graduate. In 1978 Clemens graduated at the top of his class.

Having accepted an offer to the Ph.D. program in Applied Physics at Caltech, Clemens packed his dog and his fig tree and headed to Pasadena. He joined the group of Professor William Johnson where he studied amorphous superconductors. During their initial meeting, Johnson discussed a project that involved spinodal decomposition as a step to partial crystallization. Johnson mentioned in passing that Clemens probably knew all about that, having minored in metallurgy at Colorado School of Mines. Clemens chose not to disabuse his future advisor of that belief, despite his never having previously heard of spinodal decomposition. But by their next meeting he made sure that he had found, read, and worked through all the foundational papers on the topic.

At Caltech, Clemens studied flux pinning by crystalline precipitates and compositional fluctuations in amorphous superconductors. He built Caltech's first melt spinner (a device to rapidly cool a ribbon of metal) and developed an apparatus for delivering high currents to superconductors in a high magnetic field. He modeled his critical current and field data using analytical theoretical approaches. Late in his Caltech career he met a student from Professor Marc Nicolet's group that was making amorphous materials by ion beam mixing of layered thin films. The two quickly conceived of a study of superconducting properties. Working nonstop through a weekend, Clemens did the requisite measurements and they published their findings in a paper that would lead to Clemens's first post-graduate job and to a major shift in career.

Materials science and engineering professor Bruce Clemens (left) and Jay Kadis (right) of the Stanford Center for Computer Research in Music and Acoustics measure the characteristics of a guitar for their study of guitar tone. (School of Engineering)

While in Johnson's group, Clemens—probably as a holdover from his time working in the machine shop—had the habit of coming in early in the morning. The rest of the group, including Johnson, came in late and worked late into the night. As a result, when visitors and seminar speakers came to see Johnson, Clemens was often the only group member around. This gave Clemens the opportunity to meet many famous scientists, including metallurgist Peter Haasen, and Stanford physicists Walter Harrison, Ted Geballe, and Mac Beasley, all of whom later played an important role in Clemens's career. There was a friendly rivalry between Johnson's group at Caltech and the Harvard group of Frans Spaepen and David Turnbull, whose work Clemens had first encountered when studying nucleation theory at the School of Mines. The interest these visitors showed in Clemens's work and career meant a tremendous amount to him as young graduate student and throughout his professional life.

Toward the end of his graduate career, Clemens met and married his wife, Susan. Their first son, Daniel, had the same due date as Clemens's thesis. Within one month, Clemens finished his thesis, attended the birth of his son, defended his thesis, and packed up and moved the family to Michigan to start his job in the Physics Department of General Motors Research (GMR).

Clemens's work with Nicolet on the thin film paper had led GM to hire him to develop a thin film research program. GMR was a fantastic lab, with many excellent experimental and theoretical scientists, including Gary Eesley, Wes Capehart, Jeff Buccholz, John Smith, Jack Gay, Joseph Heremans, and Jan Herbst. Clemens continued a connection with Johnson at Caltech, and worked on formation of amorphous materials by solid-state reaction of crystalline thin films. This seemingly backwards process (usually amorphous materials transform to a crystalline phase, not vice-versa) created a lot of interest and skepticism. Clemens's first talk at an international conference ended with the conference chair terminating a spirited question session with "we have to move on, but I just want to say, I don't believe it." Time showed that Clemens, Johnson, and their co-workers indeed were correct, and that this process is another way to make a metastable amorphous material.

At GM, Clemens and Eesley developed picosecond transient thermoreflectance to make the first measurements of the thermal conductivity of interfaces, and to measure the elastic modulus of thin films. They used the latter approach to demonstrate the absence of the "supermodulus effect," which had been reported (with much excitement) in multilayer films by researchers at Northwestern University and elsewhere. Clemens and Gay developed the theory of the effect of layer thickness fluctuations on multilayer x-ray diffraction, and Clemens and Professor Ron Gilgenbach at University of Michigan developed a method to form amorphous materials by self-sustained reactions initiated by rapid current pulses in multilayer films. It was a lot of fun.

Then came the fateful meeting in the sleet-soaked parking lot. Clemens finally realized that what Sinclair was suggesting was that he might apply for the position at Stanford, and so he did. In the meantime, Clemens volunteered to move to the Hughes Research Lab (HRL) in Malibu, California, in an exchange scientist program aimed at making better connections between GMR and HRL, which had just been acquired by GM. While at Hughes, Clemens worked with John Roth and Greg Olson on oxide super-conductor thin films, and also on x-ray diffraction of semiconductor superlattices, and the effect of ion beam mixing in the measurement of composition in multilayers by ion beam etch profiling. At this time, Clemens was also an adjunct professor at Caltech—where he had convinced Brent Fultz and Channing Ahn to teach him about transmission electron microscopy. At the end of his year at Hughes, Clemens was offered the position at Stanford, and in January of 1989 he started here as an assistant professor of materials science and engineering.

Clemens says that Stanford was, and is, an amazing place, with fantastic students, great colleagues, and substantial research infrastructure. Early in his career, Clemens met with Arthur Bienenstock, who invited him to consider doing research at the Stanford Synchrotron Radiation Laboratory. This proved to be transformative. Clemens and his students, working with SSRL Staff Scientist Sean Brennan, developed an apparatus for performing x-ray diffraction of thin film structure during growth by sputter deposition. Based on a suggestion by the author of this account, they also included the ability to measure the substrate curvature, and hence thin film stress, during growth. This launched a productive collaboration with the author on stresses in thin films, focusing on stress sources occurring during thin film growth. Clemens and his group also studied magnetic materials, epitaxial growth of sputtered metals, hydrogen storage materials, photovoltaic materials, and catalytic materials. The success of these efforts was partly due to the excellent collaborators, students, and post-docs with whom Clemens has had the privilege of working.

Soon after his promotion to full professor, Clemens began serving as chair of the Department of Materials Science & Engineering, a position he held from 1999 to 2004. Through a planning process led mainly by associate chair Dauskardt, Dean of Engineering Jim Plummer was convinced to give additional billets to MSE. This led to the hiring of an amazing group of faculty who have brought the department to new levels of recognition and prominence.

Clemens served as president of the Materials Research Society in 2012 and was able to experience first-hand the world-wide impact of this organization. He has served on the Faculty Senate and on several university committees, including the Safety Committee, which he chaired, during which time he co-chaired a university task force on research safety culture. He is currently chair of the Committee on Undergraduate Standards and Policy and serves as the Faculty Fellow for the Stanford Lightweight Rowing team.

Clemens is also director of the Stanford Nano Shared Facilities. This is a group of four user facilities (Nanofabrication, Electron and Ion Microscopy, Surface and X-Ray Analysis, and Soft Materials Facility) that collectively have over 1000 users, including over 800 Stanford students and post-docs, and about 200 external users from outside academic institutions and from companies large and small.

From college drop-out, to tube bender, to research scientist in a sleet-soaked parking lot, to Stanford professor, Clemens says that it has been an amazing and luck-filled journey.

Acting Assistant Professors

The traditional use of the rank of acting assistant professor has been for the appointment of junior faculty members who have not yet completed their Ph.D.'s and as a consequence are not yet eligible for regular faculty appointments. From the 1980s to 2016, however, the department used that title for young people who were appointed to serve the department as assistant professors for a limited period of time before moving elsewhere in the next stage of their careers.

As first articulated by William Tiller in 1967, the aim of the appointment was to obtain much needed help in teaching, often to make up for people on sabbaticals, and also for help with labs and the recruitment of undergraduate students. Later it was recognized that the appointments could be used to help launch the academic careers of those wishing to become professors after leaving Stanford. While the idea for this appointment had been around since the 1960s, it was not until the 1980s that the first appointment was made.

The list below indicates those former Ph.D. students who served in this capacity and the periods of their appointments. Also indicated is their first appointment after Stanford. The record indicates that the program was enormously successful in helping to launch the academic careers of many of our graduate students. Of the ten people who completed their service as acting assistant professors, seven went directly to tenure-line appointments at other universities, and another is engaged in teaching as an adjunct professor. Unfortunately, the program was terminated in 2016 on the grounds that the title of acting assistant professor should be reserved for those waiting for their regular junior appointments on this faculty.

Jeffery C. Gibeling, 1984 to 1985; professor, U.C. Davis

Bruce Lairson, 1992 to 1994; professor, Rice University

Todd Hufnagel, 1994-1996; professor, Johns Hopkins University

Rick Vinci*, 1996-1998; professor, Lehigh University

Jonathan Doan, 1998-2004; Reflectivity, TI, Solyndra, REEL Solar,

Wendelin Wright, 2004-2006; professor, Santa Clara University, Bucknell University

Seung Min Han, 2006-2010; professor, KAIST

Chris Earhart, 2010-2012; adjunct professor, Drexel

Paul Kempen, 2012-2014; post-doc., Denmark Technical University

Renee Sher, 2014-2016; professor, Wesleyan University

Ryan Brock**, 2016-2018, Exponent, Inc.

** Sadly, Rick Vinci succumbed to ALS on March 13, 2019*

*** Formally a lecturer after the termination of the acting assistant professor program.*

Chapter 20

Department Leadership and Troubles for an Aging Department

The fate of the MSE Department within the School of Engineering and the university always rests squarely on the shoulders of the people who agree to serve as department chair and associate chair. Over most of its modern history, the department has prospered though its effective leadership. Here we acknowledge the contributions provided by the eleven faculty members who have served as department chair since 1960, some of whom were on hand in 1990 to celebrate the 30th anniversary of the creation of the Materials Science Department. That celebration was followed shortly thereafter by the recognition of troubles associated with an aging department and retiring faculty, and the related difficulties associated with getting billets for replacement faculty.

Chairs of the MSE Department

Throughout the history of the department, the people who have served as its chair have had the primary responsibility for representing the department within Stanford and for looking out for the department's interests in the competition for faculty billets, space, and other resources.

Those who have served as MSE chair since 1960 are:

O. Cutler Shepard, 1960-1967

William A. Tiller, 1967-1971

John C. Shyne, 1971-1975

Richard H. Bube, 1975-1987

Stig B. Hagström, 1987-1991

William D. Nix, 1991-1996

John C. Bravman, 1996-2000

Bruce M. Clemens, 2000-2005

Robert Sinclair, 2005-2014

Paul C. McIntyre, 2014-2019

Alberto Salleo, 2019-

Celebrating 30 years of MSE

As the 25th anniversary of the creation of the Materials Science Department approached in the mid-1980s, planning began for a celebration of that event. The original plan was to have a celebration in 1985, marking 25 years since 1960. But that celebration was postponed to February of 1986 and then cancelled altogether. The reason was that the department and school were in the midst of searching for a new department chair, this time from outside of the department and school, and the faculty was not in a very celebratory mood. After Hagström was appointed as chair and before

Former chairs of the materials science department at a dinner in honor of Stig and Brita-Stina Hagström on October 12, 1991. From left to right: William Nix, Richard Bube, O. Cutler Shepard, Stig Hagström and William Tiller. (W.D. Nix)

Nix replaced him, the department did have a celebration, but this time celebrating the 30th anniversary. A modest event was held on June 1, 1990, following the final department colloquium of the academic year.

Just a year after the celebration of the 30th anniversary of the MSE Department, Hagström resigned and turned the department chairmanship over to the author of this retrospective, effective August 1991.

Hagström was thanked for his four years of service to the department and acknowledged for major changes brought about during his time. A dinner in his honor was held at the author's home on October 12, 1991.

Department chair Stig Hagström speaking at the 30th Anniversary of MSE. (W.D. Nix)

At left, Cutler Shepard, speaking at the 30th Anniversary of the MSE department. Shepard was Executive Head of the MSE department when it was founded. He is 87 years old in this photograph. (W.D. Nix)

At right, Robert Huggins, speaking at the 30th Anniversary of the MSE department. Huggins was arguably the creator of the Materials Science Department in 1960. (W.D. Nix)

At left, Richard Bube, chair of the department in the 1970s and 1980s, speaking at the 30th Anniversary of the MSE department. (W.D. Nix)

At right, John Shyne, chair of the MSE department in the early 1970s, speaking at the 30th Anniversary of the MSE department. Shyne had retired by the time of this celebration (W.D. Nix)

William Tiller, department chair in the late 1960s, speaking at the 30th Anniversary of the MSE department. (W.D. Nix)

Troubles for an Aging Department

It was during the author's term as department chair that the department went through some difficult times. Many of the faculty who had been appointed in the 1960s were starting to retire, and new faculty to replace them had not been appointed. As shown in the graph below, the number of regular faculty in MSE was stagnant during an 18-year period from 1973 to 1991. From 1977, when Robert Sinclair was appointed, to 1985 when John C. Bravman was appointed, no new appointments to the faculty were made at all. The appointments that were made during the 18-year period of stagnation were simply replacements for unexpected departures from the faculty.

Yet, the looming problems associated with the demographics of the faculty were not unanticipated. As early as 1980, one can find in the minutes of a faculty meeting the following statement: "Faculty reaching age 65 in the next ten years total zero for the department. On the other hand, 6 will reach age 65 in the four years starting in 1990, and therefore plans must start soon for that eventuality in mind." Thus, the troubles that were to come were not unexpected.

In the meantime, many faculty members continued to work largely in areas related to bulk structural materials. The department had not fully embraced microelectronics and silicon technology, topics that were dominating the research agenda elsewhere in the School of Engineering, especially in Electrical Engineering. The mindset of many faculty members in the 1960s and 1970s was that the views and opinions of people in their own research fields at other institutions around the world were more important than the views of their Stanford colleagues in other engineering disciplines.

But the author was to discover that the two-decade period of isolation from other faculty in the School of Engineering had done damage to the reputation of the department within Stanford, and that, as a consequence, obtaining approval for new faculty appointments to replace the retiring faculty was not easy. Indeed, early in my tenure as department chair, I was shocked to hear Dean James F. Gibbons suggest that the Materials Science & Engineering Department might become a division of the Department of Electrical Engineering. As might be expected, I went ballistic over that suggestion and began to put up what defense I could.

At the core of Dean Gibbons' suggestion was the fact that materials research was then being done in many different departments, including Gibbons' own department of Electrical Engineering. The counter argument that I made was that the discipline of materials science and engineering had applications all across the school and university, and that downgrading the department to a Division of Electrical Engineering would relegate Stanford to second-class status in the academic field of materials science and engineering. I pointed to top NSF and DOD fellows who were electing to go to other universities with larger and broader programs in materials science research.

To address Gibbon's point that materials science research was indeed being conducted in many different departments at Stanford, I developed an aggressive plan to dissolve the Department of Materials Science & Engineering altogether and replace it with a much larger Department of Materials Research. The proposed department would be a graduate-only department that would include many of the faculty who were then working on materials in different departments at Stanford. The plan was briefly considered by Gibbons and the chairs of the other engineering departments, but it was ultimately rejected. Two factors led to that rejection. One was that the faculty working on materials in other departments did not consider materials to be their primary professional affiliation. Instead, they considered their primary affiliations to be with such professional organizations as the Institute of Electrical and Electronic Engineers (IEEE), the American Physical Society (APS), the American Chemical Society (ACS) or the American Institute of Chemical Engineers (AIChE). Although many were active in the Materials Research Society, for example, that was not their primary affiliation. And none had any association with the more traditional materials societies of ASM International or TMS-The Minerals, Metals and Materials Society. Probably the bigger factor was that the other department chairs objected to the plan to move faculty billets out of their departments to the newly proposed department. The result was that the plan was rejected with an understanding that, going forward, the MSE department would be more active in making joint faculty appointments with other departments.

As shown in the graph below, during this hiatus the faculty head count sank to a low of

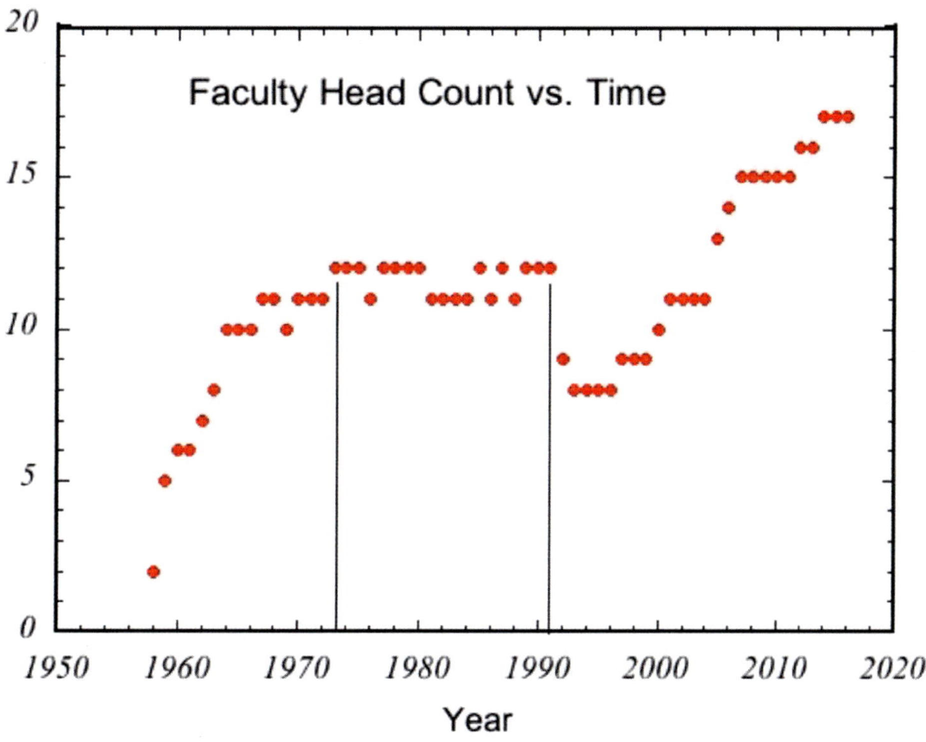

Number of regular faculty billets in the department each year from 1958 to the present.

eight, some of whom were not fully engaged in teaching and research.

Nevertheless, as faculty appointments were made, some jointly with other departments, the department's standing gradually began to improve, and, in time, Dean Gibbons became one of its strong advocates. By the late 1990s, after I had stepped down as chair, things started to improve much more dramatically, and the department entered a period of growth not seen since the 1960s. As the 100th anniversary of the department approaches, the MSE Department had made many outstanding faculty appointments, which will be described in more detail in the following chapters. Based on the work and reputation of the new faculty, the department now enjoys an excellent reputation at Stanford and is unquestionably the focal point for materials science research and education at Stanford.

Part IV ~ The Modern Department

Chapter 21

Rebuilding for the 21st Century ~ The Beginning

The early 1990s was a low point for the MSE Department—most of the senior faculty appointed in the 1960s had retired and replacement faculty had not been authorized—but the department emerged from that difficult period with the appointments of Shan X. Wang in 1993, Reinhold H. Dauskardt and Friedrich B. Prinz in 1994, and Paul McIntyre in 1997. These were the first new building blocks in what was to become the MSE department of the early 21st century. Under the 25 years of leadership of MSE chairs John Bravman, Bruce Clemens, Robert Sinclair, and Paul McIntyre, the department thrived as never before.

After Paul McIntyre was appointed in 1997, no less that nine new faculty members were hired in the 10-year period from 2000 to 2010. James D. Plummer, Frederick E. Terman Dean of Engineering from 1999 to 2014, played a big role in encouraging the rebuilding of the MSE department with an explosion of its faculty that had not been matched since the heady days of the 1960s.

Shan X. Wang ~ Magnetic Nano-technologies, Spintronics, and Biosensors

Shan Wang arrived at Stanford at the end of 1993 as an assistant professor. While he was still finishing his Ph.D. at Carnegie Mellon University, he won a junior faculty-search that had targeted information storage materials. He became the first MSE professor of Chinese descent. Born in Ningbo, China, he graduated from Zhenhai Middle School, a renowned local school, and earned the B.S. degree in physics from the University of Science and Technology of China, Hefei, China, in 1986.

Wang credited Nobel Laureate T.D. Lee's China-US Physics Examination and Application (CUSPEA) Program for affording him the opportunity to come to the US for graduate studies at a time when the Graduate Record Exam was not even available in China. He first

Shan Wang, 1993. (MSE Dept.)

obtained an M.S. degree in physics from Iowa State University in Ames, Iowa, in 1988. He earned a Ph.D. degree in electrical and computer engineering from Carnegie Mellon University, in Pittsburgh, in 1993, after studying with Professor Mark Kryder, founder and director of the Data Storage Systems Center (DSSC) there. At that time, the DSSC was regarded as the number one center of its kind in the world and to this day has remained a premier research center on information storage. Wang's expertise in both magnetic materials and devices was a perfect match for Stanford's Center for Research on Information Storage Materials (CRISM), newly established by Prof. Robert L. White. CRISM was later transformed into the Center for Magnetic Nanotechnology (CMN) under the leadership of Wang and Consulting Professor Robert M. White.

Wang began teaching classes on topics such as "Introduction to Information Storage Systems," "The Principles of Magnetic Recording," and "Electronic Behavior of Materials." He wrote a textbook, *Magnetic Information Storage Technology,* with co-author Alex Taratorin, based on his teaching and research. Later he began teaching classes in magnetism and nanostructures, and thin film synthesis. Still later, when his work turned toward biology and medicine, Wang developed a new course on biochips and medical imaging and an introductory seminar entitled "Great Inventions That Matter."

Wang initially conducted his research on inductive recording heads for hard disk drives, his doctoral research subject, and also giant magnetoresistive (GMR) and magnetic tunnel junction (MTJ) materials. The GMR and MTJ materials were the foundational materials for the burgeoning field of spintronics (also known as magneto-electronics), even before the Nobel Prize in Physics was awarded to Albert Fert and Peter Grunburg for their discovery of GMR. Around the turn of the 21st century, the Wang group (in collaboration with Professors Robert L. White and others) started to perform research in the magnetic biosensor area by applying magnetic nanoparticles (MNP) of GMR and MTJ materials to biosensing, which included DNA fingerprinting, genotyping, cancer diagnostics and cell sorting. Looking back, Wang attributed his success in the biosensor area to his experience at the Carnegie Mellon DSSC which helped condition him to try new things and pursue new areas of experimentation.

Wang's specialty in magnetism has been particularly important for medical applications because in magnetically neutral biological settings, a magnetic nanoparticle stands out like a flare in the night sky. That means if a cancer

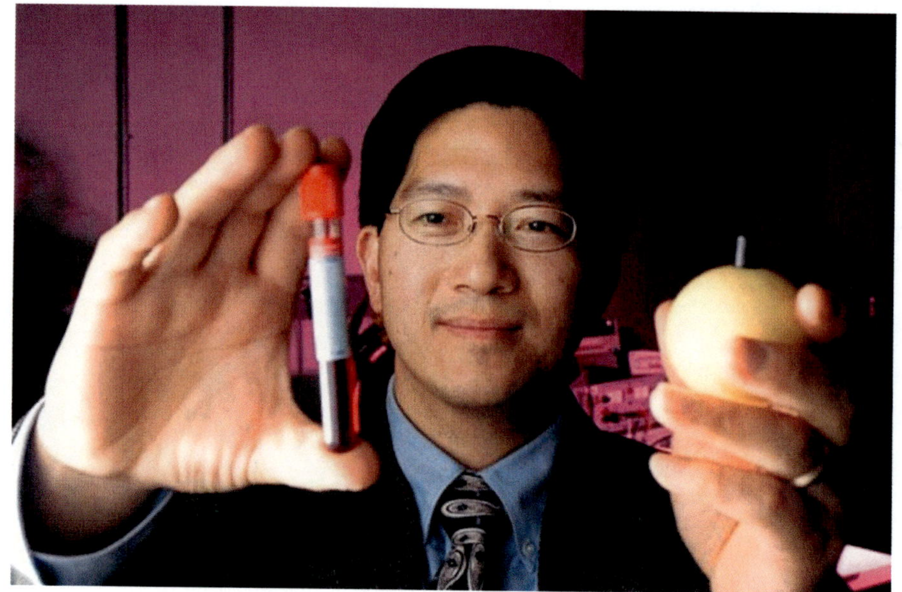

Holding a blood sample and a model of an electron with a magnetic pole to indicate its "spin," Shan Wang illustrates two applications of his magnetic nanotechnology work: fighting cancer and advancing computing, in 2006. (S. Wang)

Shan Wang and graduate student Richard Gaster demonstrate nano-sensing technology for cancer diagnostics. (Stanford News Service)

marker can be made to tag a magnetic nanoparticle, the result will be a more sensitive cancer detector than traditional signaling (like fluorescence). With better detectors, doctors can diagnose emerging cancers earlier, and can know more quickly whether a particular treatment is working. Wang trademarked MagArray biodetection chips on his own, before he had an appetite for commercialization. Many related patents soon followed and were secured by Stanford's Office of Technology Licensing. Now the biochips are the basis of two spin-off companies, MagArray Inc. and Flux Biosciences Inc. MagArray Inc. was co-founded by Wang, Stanford colleague Robert L. White, and three other professors. They spent years perfecting the magneto-nanosensor biochip technology. It finally launched a first product for early detection of lung cancer in 2018. Flux Biosciences, cofounded by Tyler Shultz and Wang, and incubated in Stanford's StartX accelerator, is developing home tests of fertility and human performance.

Wang's expertise in magnetics and his promising results in biosensors and spintronics made him a key member of two research centers created in 2006. With the support of a five-year grant of $20 million from the National Cancer Institute he co-founded, along with Sanjiv Gambhir of the department of Radiology, the Center of Cancer Nanotechnology. This Center has been renewed twice, and is currently in its 14th year of operation, having become a well-regarded center for in vitro diagnostics and in vivo imaging of early stage cancers. Also, in 2006, Wang joined the newly established Western Institute of Nanoelectronics (WIN), a spintronics research center involving Stanford and three University of California campuses. The WIN Center contributes significantly to spin-based research and development in the semiconductor industry. In the intervening years, magnetoresistive random access memory (MRAM) has emerged as an attractive candidate in new hardware architecture for fast but nonvolatile solid-state memory, energy-efficient computing and artificial intelligence (AI). Still later, in 2018, Wang joined the Joint University Microelectronics Program (JUMP), a consortium co-sponsored by the Semiconductor Research Corporation (SRC) and the Defense Advanced Research Projects Agency (DARPA). His team was tasked to develop spin-orbit torque magnetoresistive random access memory (SOT-MRAM) as a new breed of ultrafast and nonvolatile memory technology.

Wang's current research explores the use of nanosensors to detect cancer and other diseases, and he continues to be a leader in the field of information storage technology. Years ago, a group of his graduate students came up with a lab motto - "Make magnetics work for humankind, not vice versa!" - which Wang said was in part the students' tongue-in-cheek rebellion against their workload, and in part a pledge to focus on practical applications of their research.

In recognition of the work described here, in 2018 Wang was named the fifth holder of Leland T. Edwards Professorship in the School of Engineering. In naming Wang to the Chair, the Dean of Engineering, Jennifer Widom, cited Wang's contributions as being remarkable both for their impact on society and for the way they captured the imaginations of students and other faculty across the school and the university.

Wang has been recognized with numerous honors and awards since joining the Stanford faculty. He was an inaugural Fred Terman Fellow at Stanford (1994), and he was elected a Fellow of the Institute of Electrical and Electronics Engineers (IEEE, 2009) and a Fellow of American Physical Society (APS, 2012) for his seminal contributions to magnetic materials and nanosensors. His team won the Grand Challenge Exploration Award from the Gates Foundation (2010), the XCHALLENGE Distinguished Award (2014), and the Bold Epic Innovator Award from the XPRIZE Foundation (2017).

Wang and his wife, Tsing Xue, live in Portola Valley. They enjoy skiing, swimming, hiking, and travel. Their two children are now in college.

Reinhold H. Dauskardt ~ Deformation and Fracture of Materials and Interfaces

Reinhold Dauskardt, known as Reiner, joined the MSE faculty in 1994 after winning a faculty search in the area of advanced structural materials. He was the first person in 25 years to receive a tenure line appointment for work on structural materials or mechanical behavior. With Sherby and Shyne having retired in the late 1980s and this author having moved away from structural materials by that time, the MSE department had lost strength in an area for which it was once well known. And the decline came just as there was great excitement in work on advanced structural materials at some of the premier institutions with which Stanford likes to be compared (Harvard, Caltech, U.C. Berkeley, and U.C. Santa Barbara). Work on the mechanical behavior of ceramics and ceramic composites was attracting much attention in the field and the best graduate students, with their national fellowships, were migrating to these other schools. Indeed, recognition of the excitement in this area and the associated migration of students to other schools led to the search that Dauskardt won.

Dauskardt came to Stanford after having received his Ph.D. with Robert Ritchie at the University of California at Berkeley and having worked for six years as research scientist and staff scientist at the Lawrence Berkeley Laboratory. His doctoral studies came about through a joint program between the University of California at Berkeley and the University of the Witwatersrand in South Africa, where he had started his graduate work. At Berkeley, he had discovered that ceramics, ceramic composites, and other brittle materials were susceptible to mechanical failure by fatigue, a failure mode almost always associated with ductile structural metals. He brought his expertise in fracture and fatigue to the department and later used it not only to continue his work on structural ceramics but also to initiate new work on bulk metallic glasses and the adhesion of microelectronic thin film interfaces.

He taught the courses on "Mechanical Behavior of Materials," and introduced a new course on "Fracture and Fatigue of Materials," later updated and expanded to include "Thin-Film Structures." Coursework on fracture and fatigue had been offered only sporadically since the 1960s, when Tetelman was on the faculty. Dauskardt also created a new course entitled: "Microstructural Design of Advanced Materials and Composites." Altogether, through his teaching and research, Dauskardt reintroduced mechanical properties of materials to the department.

Reiner Dauskardt was born in Johannesburg, South Africa, on July 31, 1959. His German-born father trained as a watchmaker there, and migrated to South Africa in the 1950s. His mother, whose family had been in South Africa for several generations, is also of German descent. She works closely with his father and looks after the administrative side of their long-standing watchmaking and repair business. The elder Dauskardts currently live in the Stanford area and have clients from all over the US.

Dauskardt was schooled close to Johannesburg in Germiston, South Africa. He graduated from the JDK Technical High school in Germiston before enrolling at the University of Witwatersrand (called "Wits") in Johannesburg, to study physics and mathematics. He received his B.S. degree from Wits in 1980 and continued working at Wits toward an M.S. degree. During his time there, he had the rare treat to study with Professor Frank Nabarro, one of the giants of metal physics, as one part of the field of materials science was once called. Dauskardt learned thermodynamics from Nabarro, but mostly was taught to be on his toes when it came to scientific discourse. Nabarro used the Socratic method in his teaching and lectures, during which he frequently peppered his students (anyone listening was a student) with questions about what should come next in the argument. In his several lectures at Stanford, Nabarro used the same technique and expected senior faculty to be prepared to respond if asked. Dauskardt stayed in contact with Nabarro throughout the rest of Nabarro's life, which included his several visits to Stanford, the last just a year before he died. As Dauskardt says: "Having dinner with Nabarro was like having an oral exam, with him asking probing questions on all aspects of materials physics."

Reinhold H. Dauskardt (MSE Dept.)

Dauskardt's graduate work at Wits started in the Department of Metallurgy from which he received an M.S. degree in 1983. He then enrolled in the joint Ph.D. program between Wits and U.C. Berkeley and completed his degree with Robert Ritchie in 1988.

At Stanford, Dauskardt has made pioneering contributions to the study of thermomechanical and fracture behavior of materials, thin-films and soft tissues that have had fundamental and technological impact. At first, he continued the work on fracture and fatigue of ceramics and glasses he had started at Berkeley and focused primarily on the effects of moisture on these crack growth processes. Then, in the late 1990s, he turned his attention to the fracture of bulk metallic glasses. After the bulk metallic glass Vitreloy 1 was developed in the early 1990s, there was great excitement about using these exceptionally strong materials in structural applications. While the new materials were known to have no useful tensile ductility, their fracture properties had not been systematically studied. In the late 1990s and early 2000s, Dauskardt and his students published a series of papers that laid the foundation for understanding fracture and fatigue of bulk metallic glasses and the damage processes that occur at crack tips in these materials during failure. This body of work includes some of Dauskardt's most highly cited papers.

By the 2000s, Dauskardt was turning more and more of his attention to the failure of interfaces in thin film devices. He made important contributions to developing quantitative metrologies for characterizing adhesion in thin-film structures and linking adhesive properties to underlying materials chemistry, process variables and structure using analytical and computational models. Dauskardt's work on developing quantitative adhesion metrologies has had a significant impact on manufacturing process yield, device design, new materials integration, and in-service reliability of thin-film semiconductor, solar energy and other thin-film device and coating technologies. He has helped more than 100 international device companies establish adhesion characterization capabilities based on tools and methods that he developed. The impact of this work on the thermomechanical reliability of devices was recognized by the prestigious Semiconductor Industry Association University Researcher Award in 2010 for "research which has provided substantive and sustained contributions to semiconductor industry science and technology."

Reinhold Dauskardt, with student Dan Maidenberg, 2006. (School of Engineering)

Dauskardt has also extended his work on adhesion to soft tissues, including human skin. There he developed the first quantitative biomechanics framework to characterize, understand and control the biological processes that determine the effects of treatments and exposures on the sensory perception, firmness and damage processes in human skin. His work on regenerative processes and scar formation in healing cutaneous wounds has laid the groundwork for new and clinically proven technologies for controlling hypertrophic scar formation in humans. His work was recognized with the 2011 Henry Maso Award from the International Federation of Societies of Cosmetic Chemists for advancing the understanding of the role of treatments and exposures. He is co-inventor and co-founder of NeoDyne Biosciences, a company that has the first clinically proven, patented, and US FDA approved scar reduction technology for humans.

Dauskardt has contributed significantly to leadership in the materials community at Stanford. He served as Associate Chair of MSE for 15 years from 2000-2014. He has also contributed significantly to international leadership in multidisciplinary materials and bioengineering efforts at Stanford. Since 2016 he has been a Visiting Professor in the School of Materials Science and Engineering at Nanyang Technological University in Singapore.

His interdisciplinary research includes interaction with researchers in academia, industry, and clinical practice. He has published over 300 articles in the scientific literature and holds more than 25 U.S. and International Patents and has delivered over 200 invited and plenary talks and lectures. Dauskardt has contributed significantly to leadership in Stanford's multidisciplinary materials efforts, which span several departments, schools, and laboratories. He serves on several international agency and company advisory boards. He has supervised 50 doctoral students and 34 postdoctoral fellows, medical residents and fellows, and currently has 14 doctoral students and 6 post docs in his research group.

In addition to the awards already cited, Dauskardt has received an Alexander von Humboldt Research Award; the Structural Materials Distinguished/Engineer Award from the Metallurgical Society for "Long lasting contributions to the fundamental understanding of microstructure, properties and performance of structural materials for industrial applications, along with dedication and leadership of the society;" and The Semiconductor Industry Association University Research Award for "Research which has provided substantive and sustained contributions to semiconductor industry science and technology." He was elected Fellow of both ASM International and the American Ceramics Society. He was appointed the Ruth G. and William K. Bowes Professor of Engineering in 2013.

Paul C. McIntyre ~ Advanced Integrated Circuit Materials and Leadership at Stanford

Paul McIntyre joined the MSE faculty in 1997 after receiving his Sc.D. degree from MIT in 1993 and having worked as a post-doc at the Los Alamos National Laboratory and as a staff member at Texas Instruments. The department had decided to conduct a search for a faculty member in the broad area of ceramics, thinking that an appointment might be made in structural ceramics, a hot field of research at the time. Several people working in structural ceramics were considered for the appointment, but when McIntyre interviewed for the position and described his then-current work on gate dielectrics at Texas Instruments and his earlier work on superconducting ceramics at MIT, members of the search committee quickly revised their thinking. During the interview process, one senior faculty member who had spoken only briefly to McIntyre marched into the Chair's office and said: "Hire this person immediately," so taken was he of the depth of McIntyre's work and his suitability for a faculty appointment.

McIntyre developed a reputation as that member of the MSE faculty who would do the best in taking the tough oral qualifying examination, which at the time still comprehensively, (to a grueling degree) covered all aspects of materials science. He was said to have the widest "bandwidth" of all members of the faculty. That depth of knowledge and experience served him well in his teaching. In addition to introducing a new course on electronic materials and a freshman chemistry course for engineers, he became one of the stalwarts of teaching the fundamental core courses in thermodynamics, phase equilibria and kinetics. In research, he brought the development of high capacity gate dielectric oxides to a new level and introduced research on semiconductor nanowires for transistor applications. He has also assumed leadership roles at Stanford by serving as director of the Geballe Laboratory for Advanced Materials, a Stanford independent laboratory devoted to interdisciplinary materials research (2008-2013), as chair of the Materials Science and Engineering Department (2014-2019), and as director of the Stanford Synchrotron Radiation Lightsource (2019-).

McIntyre was born in Vancouver, British Columbia, Canada, on September 21, 1965. His father was an accountant who worked as the treasurer and, later, the finance director of a local municipality. Before her marriage, his mother had graduated from the University of British Columbia and then worked as a librarian in the Vancouver Public Library system. She became a homemaker, eventually having three sons (McIntyre was the oldest), before returning to the workforce as a secretary.

Paul C. McIntyre, 2017. (School of Engineering)

McIntyre grew up in what was then a truly rural part of British Columbia, in and around the Fraser Valley towns of Abbotsford and Aldergrove. For several years, he lived near an Indian reservation and became friends with several of the Indian kids. They introduced him and his brothers to the best spots for trout fishing in the local ponds and streams. McIntyre's family had a farm there with about 40 head of beef cattle (mainly Hereford) about 50 hens; several pigs; and an extensive vegetable garden and orchard. In addition to less predictable activities, such as mending fences, digging post-holes, baling and stacking hay bales, de-horning and castrating cattle, and plucking freshly killed chickens, McIntyre had regular chores on the farm. His father owned several American Saddlebred horses, which he would take to horse shows in BC and the Pacific Northwest states of the US. It was McIntyre's job to feed the horses and clean their stalls each day, which he did from the time he was about 11 years old until he left home to attend college.

McIntyre attended Abbotsford Senior Secondary School in Abbotsford, British Columbia, graduating summa cum laude in 1983. His favorite subjects were physics, mathematics, chemistry and western civilization. He wasn't much of an athlete, although he enjoyed playing pick-up games of ball hockey, soccer, and tennis with friends.

By the time McIntyre had graduated from high school, he was ready to leave life on the farm behind! He attended the University of British Columbia in Vancouver, which at the time seemed like the big city to him. He decided to study engineering because it put to practical use the science subjects he had so enjoyed studying in secondary school. Also, he had grown up watching the Apollo moon landings, the launch of space probes like the Voyager and Viking interplanetary missions, and he had witnessed the arrival of the first personal computers, home video recorders, and digital recording of music. He was excited, he recalls, about engineering, because it was bringing all these new technologies into peoples' lives and allowing humanity to do great things.

At that time, UBC had a five-year program to earn a bachelor's degree in applied science (engineering). One took a first year of general science courses and had to obtain a high GPA to be allowed to progress into the engineering curriculum itself. At the end of the next year, which would be the student's first year in engineering, he or she would choose a major. Having done well in his year of general science at UBC, McIntyre entered first year engineering and but was disappointed with many of the courses he took that year. He was bored with statics and dynamics, and with engineering drawing, engineering economics, and statistics. He liked thermodynamics better, but his favorite course that year, hands-down, was engineering materials. He appreciated the fact that this subject was firmly rooted in physics and chemistry but was very relevant to engineering practice. The class taught him a little bit about crystalline structures of solids, structure-property relationships, and thermodynamics of phase changes—which he considered very exciting stuff! It was probably the biggest factor in his choice to major in Metals and Materials Engineering. He earned his

BASc degree in that major in 1988, then left Canada for Boston.

McIntyre attended MIT, where he earned his doctor of science (Sc.D.) degree from the Department of Materials Science and Engineering in the field of ceramics. At MIT, doctoral students in the School of Engineering can select either the Ph.D. or the Sc.D. for their degree. The degree requirements are identical. Most students choose the Ph.D. because it's familiar and recognizable to prospective employers. McIntyre chose the Sc.D. precisely because it was different and it was an "MIT thing," something not available at most other schools.

At the time, superconducting ceramic metal oxides were attracting great attention both in the condensed matter physics community and among engineers. In his thesis research, under the supervision of Michael Cima, he developed a non-vacuum method for depositing films of the high-Tc superconducting compound yttrium barium cuprate (YBCO), using metalorganic precursors dissolved in an organic solvent. This process, which he and Cima patented, was later commercialized by the company American Superconductor, and was used in their superconducting tape manufacturing business. In his thesis project, he identified the mechanisms by which the deposited metalorganic films were transformed, through an amorphous metal oxyfluoride intermediate phase, to superconducting YBCO. He discovered that this involved transient melting of the oxyfluoride intermediate, and that this transient melting could be used to achieve liquid phase epitaxial growth of large-area single crystal YBCO films on appropriately chosen substrates.

McIntyre stayed on at MIT for almost a year after graduation, to write up papers and to perform additional experiments inspired by the results obtained in his doctoral thesis project. In April 1994, he moved to Los Alamos National Laboratory as a Lab Director's-Funded Postdoctoral Fellow. His principal advisor at LANL was Michael Nastasi. His postdoctoral research focused on epitaxy of metals on single crystal metal oxides, characterization and energetics of crystalline interfaces, interdiffusion in metal multilayers and ion beam materials analysis.

In August 1995, he joined the Central Research Laboratories of Texas Instruments (TI) as a member of the technical staff. He worked on a large, DARPA-funded project that also involved IBM, Micron Technologies and university researchers in the US and Germany. The project focused on integration of new, ultra-high permittivity dielectric materials in silicon microelectronics. Their goal was to replace amorphous silicon dioxide and silicon nitride, industry-standard dielectric materials, with barium strontium titanate, a crystalline dielectric alloy based on the bulk ferroelectric material, barium titanate, and to use this dielectric in the on-chip capacitors that are an essential part of DRAM memory. At the same time, he collaborated with a small team of researchers at TI who were doing some of the very first studies in the world on replacing silicon oxide dielectrics in field effect transistors with deposited metal oxide dielectrics. This is a line of research that he quickly adopted after coming to Stanford.

McIntyre joined the faculty of the Materials Science and Engineering Department at Stanford in January 1997. Among the first things he noticed about the department was that, even as an assistant professor, his more senior colleagues wanted to hear his opinions on important questions related to the direction of the department. He benefited tremendously from the advice and mentorship of many faculty, particularly William Nix and David Barnett.

One interesting piece of advice was provided by Oleg Sherby, shortly after McIntyre arrived at Stanford. Sherby told him that, prior to his retirement, he had taught the same graduate course in mechanical behavior of materials for many years. He had performed an experiment in which, during successive years offering the course, he would either wear a necktie for every lecture or he would teach every lecture wearing a dress shirt with no necktie and an open collar. Sherby claimed that there was a statistically significant increase in his teaching evaluations when he taught while wearing a necktie, and he strongly advised McIntyre to do this whenever he gave a lecture! McIntyre was more than a little skeptical; after all, he had just joined the university after working in an industrial R&D lab where none of the staff wore ties, unless a VP or some other potentate was visiting. It seemed absurd to don a necktie to teach students who, even then, were a good deal younger than he. He did not take Sherby's advice. As he says, "considering the teaching evaluations earned in my first quarter as an instructor at Stanford, perhaps I should have!"

Over the years, McIntyre developed a new version of the undergraduate electronic materials course, MatSci 152, and a freshman chemistry course intended for prospective engineers, ENGR 31. For about 10 years, he taught MatSci 204, the department's graduate core course on phase equilibria and thermodynamics. During the past 10 years, he has taught MatSci 207, the core course on kinetics.

At Stanford, McIntyre established a research group that focused initially on semiconductors and related materials required for development of new microelectronic devices and for the continuation of Moore's Law. Given his experience at TI, the primary focus of his initial work was to understand how to integrate new metal oxide materials for capacitive charge storage and transistor gate dielectrics on silicon. Over the years, his group developed important capabilities in a film deposition method called atomic layer deposition, a pulsed form of chemical vapor deposition, and they used it prepare and study ultra-thin metal oxide and metal nitride films. They became increasingly involved in combining advanced spectroscopy methods, such as synchrotron-based photoelectron spectroscopy, with electrical characterization of bulk and interface defects in transistors, and they developed methods to passivate such defects in many different oxide/semiconductor systems.

Much of this work was done in collaboration with Krishna Saraswat and Yoshio Nishi in Electrical Engineering. They also started a major activity in growth, characterization and photonic applications of semiconductor nanowires, particularly group IV (silicon-germanium-tin) nanowires. Finally, in collaboration with Christopher Chidsey in Chemistry, they used atomic layer deposition to synthesize protective and transparent coatings that allow silicon solar cells to operate stably in contact with water. Some of their ALD-derived coatings also can catalyze electrochemical reactions such as water oxidation. This has allowed them to build efficient (> 10% solar-to-hydrogen efficiency), integrated silicon water splitting cells for solar-driven hydrogen synthesis, a promising approach to cope with the intermittency of solar power by storing some portion of the collected solar energy in the form of chemical bonds.

McIntyre has received a number of honors and awards during his time at Stanford. He is currently the Rick and Melinda Reed Professor in the School of Engineering and Senior Fellow, by courtesy, at the Precourt Institute for Energy. He was awarded the Woody White Service Award by the Materials Research Society in 2011.

Chapter 22

Rebuilding for the 21st Century ~ The Explosion

Although the rebuilding of the department for the 21st century had begun in the 1990s, the year 2000 marked the beginning of an explosion in the size of the MSE faculty leading to an increased number of undergraduate students studying materials science. From that point forward, the Department of Materials Science and Engineering became the focal point for materials science teaching and research for the university.

Several factors contributed to the resurgence of the MSE department at this time. One was that department had begun to make faculty appointments jointly with other engineering departments, as it had been strongly encouraged to do. Another was that some other engineering departments that had been dominating solid-state materials research at Stanford, most notably Electrical Engineering, developed other non-materials interests, namely software and computing in the case of EE.

Still another factor was that the MSE department began to select faculty members on the basis of their work on the applications of materials to important technological problems, rather than following past practice of selecting individuals with a focus on one particular aspect of fundamental materials science. Recall that the name of the department from 1961 to 1971, when most of the initial growth of the faculty occurred, was Materials Science; the "and Engineering" had not yet been added. It is not surprising that most of the faculty appointed during that time focused on the fundamental science of materials, and not on engineering applications. For example, William Tiller was appointed based on his work on the science of solidification and initially worked in that area. Oleg Sherby was appointed based on his work in understanding the mechanisms of high temperature creep of metals and he worked in that area throughout his career. Marshall Pound worked on the theory of nucleation, and while Richard Bube worked on solar cell applications later in his career, he was engaged during the 1960s mainly in fundamental work on photoelectronic properties of II-VI compounds.

By the turn of the 21st century, however, well after the name of the department had been changed to Materials Science and Engineering, most faculty in MSE were making their reputations by focusing on what one can do with materials in engineering applications rather than how one can manipulate their properties or behavior by changing their structure though processing. Examples include the development of solar cells, batteries, smart windows, chemical and biological sensors, electronic, photonic and plasmonic devices and stem cell therapies. Thus, the focus was less on how materials were constituted and how they behaved, than on how material behavior could be used to enable certain technological developments. Developing a more efficient solar cell, a new battery system, or a new device based on plasmonics became more important than developing an ever more sophisticated theory or model of material behavior. Another related shift was the move toward studying electronic, optical, and magnetic properties of materials and away from the mechanical properties that had been the focus of the department in earlier times.

The changing emphasis of the department later led to an increased number of undergraduate students who were more interested in engaging in efforts to solve bigger societal problems than in perfecting our understanding of material behavior. The graph below shows that around 2008 there was a significant up-tick in the number of students graduating from the department with the B.S. degree. That dramatic increase correlates with the increased numbers of new faculty appointed earlier in that decade. The new faculty felt strongly that more attention should be focused on undergraduate students and they created more courses of interest to undergraduates. The result was that by the second decade of the 21st century, the rate of granting B.S. degrees was higher than it had ever been before. This is in stark contrast to the period from the late 1960s to the late 1970s, when almost no B.S. degrees were granted, and when almost all attention was focused on specialized, often theoretical, research of interest to graduate students only.

Michael McGehee ~ Polymeric Solar Cells and Sustainable Energy

Michael McGehee was one of the first of a new breed of faculty members, appointed around the turn of the 21st century, who were

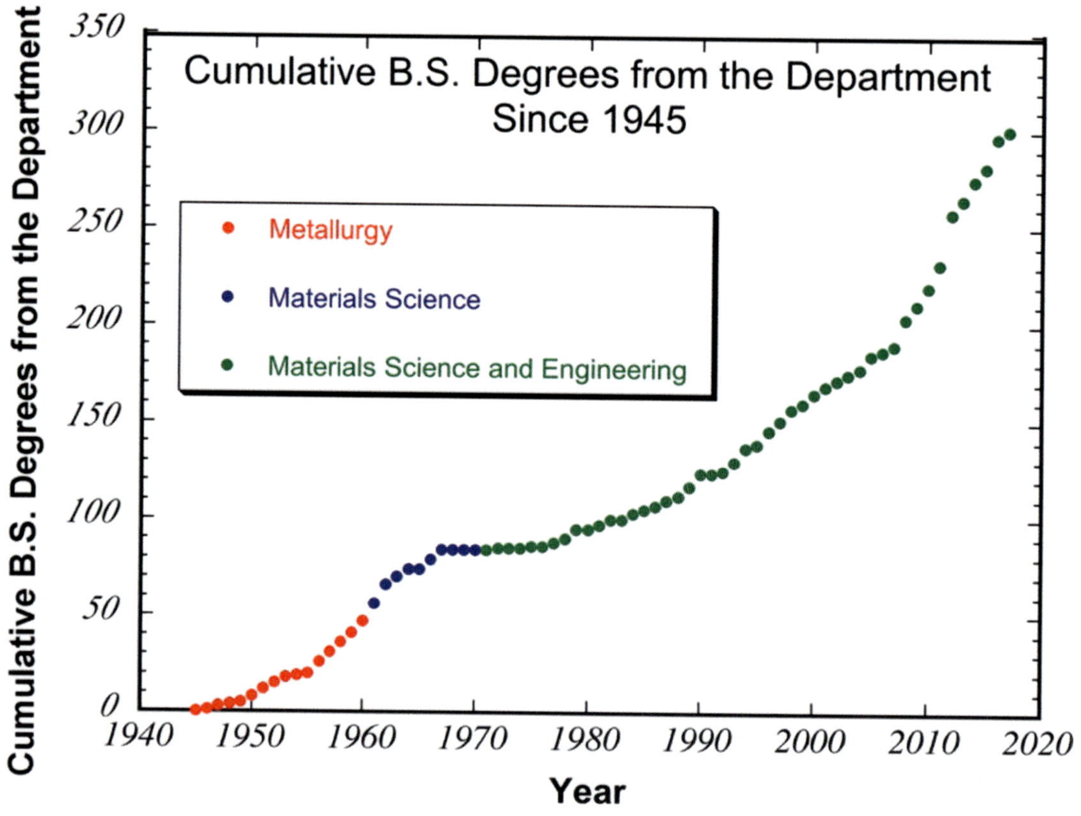

Cumulative B.S degrees granted by the department. Notice the lull in the 1970s and the rapid rise around 2005.

more interested in using materials principles to make new materials to solve major societal problems than in studying the principles themselves. As Robert Huggins has reported, the School of Engineering's dean in the 1960s, Joseph M. Pettit, was frequently critical of the department for its focus on principles and theory. He would say of the department, "You don't make anything." With the appointment of McGehee and others, that all changed.

McGehee had worked with Alan J. Heeger at UCSB on the development of light-emitting polymeric semiconductors that could be used to make solid-state lasers. These were attractive engineering materials because they could be processed easily by spin casting and did not require the perfection and cleanliness needed for semiconductor processing with silicon or related semiconductor compounds. McGehee brought that approach with him to Stanford and was soon making polymeric solar cells with impressive efficiencies. During his time on the faculty, essentially all of his research was focused on topics related to organic solar cells, field effect transistors, light-emitting diodes, and the lasers or transparent electrodes needed for those devices.

McGehee's teaching was closely related to his research interests. During his time in the department, he taught courses in nano-technology, organic semiconductors, organic materials, solar cells and nanocharacterization. He viewed teaching more as a way to expose students to important societal problems that could be addressed with materials science and less as a way to deepen a student's knowledge in a particular branch of the field or to test her/his ability to solve classical problems in the field. His teaching put an emphasis on a student's knowledge of the outstanding problems that needed to be solved. While those with a more traditional orientation might have questioned this approach to teaching, they could not quibble with McGehee's success in preparing his students for productive careers in the field. His students have started a dozen

Michael D. McGehee (Michael McGehee)

companies and six have won the Forbes 30 under 30 Award.

McGehee's most highly cited paper, "Conjugated polymer photovoltaic cells," published in *Chemistry of Materials* in 2004, was co-authored with one of his first doctoral students, Kevin Coakley. By late 2018, that paper had been cited 1816 times according to the ISI Web-of-Science website. That paper, and ones that followed soon thereafter, brought McGehee significant national acclaim and led to many awards, including his MRS Outstanding Young Investigator Award, mentioned elsewhere in this document. With Joe Kline, another of his first doctoral students, McGehee wrote other highly cited papers on conjugated polymeric semiconductors for use in field effect transistors. These papers, and the hundreds more that would follow during his time at Stanford, would lead to one of the most outstanding publication records by an MSE faculty member. By 2018, McGehee had a Web-of-Science h-index of 85 (85 publications cited at least 85 times) with an average of 106 citations for each publication.

Later in his Stanford career, McGehee conducted research on perovskite solar cells. That research was built on his prior investigations of organic light-emitting diodes and lasers, light extraction from light emitting diodes, light trapping in solar cells, and charge transport in organic semiconductors. He was inspired in this and previous work to develop technology that can provide humanity with clean energy and can solve environmental problems. Although his projects had clear applications involving close interactions with companies, he still took a fundamental scientific approach that involved sophisticated characterization and advanced modeling.

In related work, he engaged in research on what he called "Dynamic Windows." He cited a tremendous opportunity to use windows with adjustable tinting to improve the energy efficiency of buildings and to provide people with a more pleasing environment with optimized natural lighting. Although there was already a commercial market for electrochromic windows, the market had not yet exploded for a variety of technical (color, switching speed) and economic reasons. McGehee's approach was based on reversibly electroplating metal onto transparent electrodes. His glazings had a very neutral color and had excellent potential to be more cost-effective than previous commercial electrochromic systems. At the time of this writing, he is developing methods for improving the conductivity of his transparent electrodes, which are expected to enable large windows to switch modes in under 30 seconds.

In addition to winning the MRS Outstanding Young Investigator Award, McGehee received numerous other awards and recognitions for his work. He received the Vance and Arlene Coffman Faculty Scholar Award (2007), the Mohr Davidow Venture Innovators Award (2007), and the Kosuke Ishii Award for Industry Education Innovation (2012), among many others. He was the Gilbreth Lecturer at the National Academy of Engineering's Meeting (2006) and the Global Climate and Energy Project Distinguished Lecturer (2012). He also served as director of the Center for Advanced Molecular Photovoltaics at Stanford from 2008 to 2012.

Michael D. McGehee was born in Richmond, Virginia, on October 25, 1972. His father was the vice president of a propane company while his mother ran a pre-school. He was raised on the outskirts of Richmond, and received his primary and secondary schooling there. He graduated Douglas Southall Freeman High School in Henrico, Virginia, in 1990 and enrolled that fall at Princeton University, where he majored in physics. After graduating from Princeton in 1994, he moved to California for graduate work in the Materials Department at UC Santa Barbara. There he pursued his Ph.D. degree by working on semiconducting polymers for solid-state lasers with Alan J. Heeger as his advisor. (In 2000, Heeger would share the Nobel prize in Chemistry with Alan G. MacDiarmid and Hideki Shirakawa for their "discovery and development of conductive polymers.") McGehee graduated with his Ph.D. in 1999, and subsequently worked as a post-doctoral student at UCSB with Galen Stucky for a few months before accepting an appointment, in 2000, as an assistant professor in the MSE Department at Stanford.

McGehee left Stanford in 2018 to accept an appointment as professor of chemical and biological engineering at the University of Colorado, Boulder, with a joint appointment at the National Renewable Energy Laboratory. He moved to Colorado, in part, for a better lifestyle with easy access to good hiking and affordable housing.

Mark Brongersma ~ Nanoscale Photonics

Mark Brongersma joined the MSE faculty in 2001 as an assistant professor after having completed post-doctoral work at Caltech with Professor Harry Atwater in the area of nanophotonics. Brongersma was the second MSE faculty member to have come from Caltech (Clemens was the first), and he opened the way for five more Caltech students to join the MSE faculty. It was at Caltech that Brongersma

Mark Brongersma, 2005.
(Stanford News Service)

worked on and coined the word "Plasmonics" for a new technology in which light is transported and manipulated with metallic nanostructures. He has continued to work in that area at Stanford where his research is directed towards the development and physical analysis of nanostructured materials that find application in nanoscale electronic and photonic devices. He is widely known both nationally and internationally and has received major awards for that work. He also became a very effective teacher, winning Stanford's prestigious Walter J. Gores Award for Excellence in Teaching, in 2007.

Brongersma was born in the town of Geldrop in the Netherlands on October 29th, 1969. He comes from a very distinguished family of scientists. His father, Hidde Brongersma, first worked for the Philips Electronics company and later as a professor at the Technical University of Eindhoven. He currently is the CEO of the startup Calipso and an honorary professor in the Materials Department at Imperial College in London, England. His mother, Ulrika Brongersma, is a Ph.D. chemist, but ultimately gave up her scientific career to raise Mark and his brother Sywert, also a well-known Ph.D. materials scientist, and to enjoy gardening, playing the cello, and, as Brongersma has said, "all those other things that great mom's do." His extended family includes many other scientists; all four grandparents were also Ph.D. scientists (and three were university faculty members). As such, early on he was caught up in the exciting world of academia.

Brongersma went to high school at the Hertog Jan College in Valkenswaard, The Netherlands, from 1984 to 1990. He very much enjoyed his mathematics, chemistry, and physics classes, but also gained an interest in art history and played classical guitar. After high school, he enrolled at the local Technical University of Eindhoven to study physics, where he received his M.S. degree in August 1994. During his studies there, in 1992, he participated in a research project at AT&T Bell Laboratories in Murray Hill, New Jersey with Dr. Walter Brown. Brown was very influential on Brongersma's scientific thinking and took significant time to teach him about ion-solid interactions and ways to approach scientific challenges. Brongersma also did a project at the FOM-Institute in the Netherlands on the synthesis of silicon (Si) nanocrystals. He was so excited about this research and the energetic atmosphere at the

Institute that he decided to stay there for his Ph.D. research.

For his Ph.D., Brongersma worked under the supervision of Alberto Polman. He started with projects aimed at understanding the flow of glasses under the irradiation of a high-energy ion beam. Later, he developed ways to use ion beams to create light-emitting Si nanoparticles. Through this project, he got increasingly interested in the optical properties of quantum emitters. This research also led to a very fruitful collaboration with Professor Harry Atwater at the California Institute of Technology (Caltech).

After finishing his Ph.D., in 1998, Brongersma left for California to join Harry Atwater's Caltech group and started the first nanophotonics project in his group. As noted, there he worked on and coined the word "Plasmonics" for a new technology in which light is manipulated with metallic nano-structures. After two exciting and productive years at Caltech, Brongersma started interviewing for jobs in the Netherlands. His aspirations to go back to the Netherlands all changed when he met Kathy Dizio, his future wife, at Caltech. She had driven across the U.S. (with a motorcycle and horse in a horse trailer) to join Professor David Tirrell's group. Interestingly, she got to work with Sarah Heilshorn in the Tirrell group and at that time Nick Melosh and Andy Spakowitz (all future MSE faculty members) were also local friends.

Mark Brongersma 2016. (M. Bronsgersma)

Soon after they met, Brongersma and Dizio moved in together in a place that borders the Natural Forest, in Altadena, and all his plans to go back home got cancelled. Instead, Kathy suggested that Brongersma should instead try to get a job at Stanford, because her mother lived near Stanford. Somehow all the stars were aligned and Brongersma was offered a position in the MSE department (luckily, as he naïvely had not applied to any other universities.). Before coming to Stanford, he also learned about cavity quantum electrodynamics while spending a short year in Professor Kerry Vahala's group at Caltech.

Brongersma enjoys teaching a diverse set of courses at Stanford, ranging from surface micro-analytic techniques, to semiconductor device physics, to nanoscale photonics. His passion for teaching was rewarded with the Walter J. Gores Award for Excellence in Teaching, in 2007.

At Stanford, the Brongersma group initially worked on many aspects of manipulating light at the nanoscale. Early on his research was focused on the development of nanoscale light sources, modulators, and detectors. He also continued research on Plasmonics and for his research on this topic was awarded the Raymond and Beverly Sackler Prize in the Physical Sciences for Physics in Israel in 2010. More recently, the group has changed its focus to the development of metamaterials that can be used to replace the bulky optical components we find in imaging and augmented/virtual reality systems. In this work the development and understanding of new optical materials is a key aspect. His current research includes work on plasmonics, semiconductor nanophotonics, microcavity resonators and infrared nanooptics.

Brongersma is very active in the materials, physics, and photonics communities and is a

Mark Brongersma with students in his lab, 2006. (School of Engineering)

Fellow of the Optical Society of America since 2008, a Fellow of the American Physical Society since 2010, and a Fellow of the SPIE (an international professional society for optics and photonics technology) since 2011.

Nicholas A. Melosh ~ Inorganic Structures in Biology and Energy Conversion

Nicholas Melosh received his B.S. degree in chemistry from Harvey Mudd College in 1996 and went on to do a Ph.D. in Materials Science at UC Santa Barbara working with Brad Chmelka, Galen Stucky, and Glenn Fredrickson. He then worked with Jim Heath at UCLA and Caltech as a post-doc from 2001 to 2003 before joining the Materials Science & Engineering Department at Stanford in 2003. His interests include interfacing inorganic structures with biology, unconventional energy conversion, and plasmonics. He currently is a Terman Fellow and the Reid and Polly Anderson Faculty Scholar at Stanford University.

Melosh was born in Chicago in 1973 but lived for a time in Pasadena and on Long Island before arriving in Tucson, Arizona, in the mid-1980s. His father, H. Jay Melosh, a geophysicist, had served on the faculties of Caltech and the State University of New York before joining the faculty at the University of Arizona. (He later moved to Purdue University where he is a University Distinguished Professor in the department of Earth, Atmospheric and Planetary Sciences.) Nicholas Melosh attended University High School in Tucson, graduating in 1992. He enrolled at Harvey Mudd College that year and eventually studied at UCSB, UCLA, and Caltech before arriving at Stanford.

After starting his undergraduate career in chemistry, Melosh wanted to push into an area with greater real-world impact, so he enrolled in Materials Science at UCSB for his Ph.D. degree. During that time, UCSB was one of the world leaders in polymers and polymer physics, so Melosh started studying block co-polymer self-assembly. There he used the local structure of the polymer species to direct the partitioning and solidification of an inorganic species, in this case silica glass. This created hybrid structures with the inorganic phase patterned by the organic species, producing highly ordered and high surface area materials for catalysis and optoelectronics.

Following his Ph.D. work, Melosh shifted gears and went to Jim Heath's lab at UCLA (soon to move to Caltech). The focus there was quite different; it was about using complementary metal–oxide–semiconductor-type (CMOS-type) processing to make molecular electronic

Nicholas A. Melosh (MSE Dept.)

devices for nanoscale circuits. Melosh learned nanoscale and cleanroom fabrication and developed a process to make wires and circuits at the 8nm level, at the time unheard of in terms of resolution and density.

Starting in the fall of 2003 at Stanford, Melosh's initial plan was to pursue combinations of polymer-like self-assembly together with electron transfer for molecular electronics applications. The goal was to use molecular species to enable scaling of electronic devices beyond the 100-nm "limit." While laughable today in a worldwide era of 14 nm wires in smartphones, at that time this was still thought to be a difficult physical barrier. He started working with Mark Brongersma, who had also recently joined the department, and they soon developed a new opto—electronic method of interrogating the minute changes in these types of molecular species under electric fields, switching them from one form to another. These results largely revealed that the much-hyped molecules in molecular electronics did not work nearly as well as commonly thought. At the same time, the reverberations of the Hendrik Schön scandal at Bell Labs (which involved very similar molecular electronic devices) raced through the research community, nearly wiping the field off the scientific map.

At that point, as an untenured assistant professor, Melosh had to change the direction of his group. Taking the skillsets from his Ph.D. and post-doc, he re-directed half of his group to focus on molecular self-assembly at inorganic interfaces; they eventually became the 'bio' side of his team, while the other half started to look more deeply at electron transfer across interfaces, focusing on a new type of molecule, diamondoids, with an emphasis on energy applications.

The advent of the discovery of 'diamondoids' kicked off a new avenue of intellectual inquiry. Extracted from refined petroleum, these are small molecular fragments of diamond generally 1-5 unit cells in size, and they are a bridge between small molecules, long the domain of chemistry, and nanoparticles, which is where materials science often starts out. Diamondoid molecules have a wide variety of shapes and sizes, yet were all from the same diamond family, allowing detailed examination of the transition between molecules to nanoparticles.

Since diamondoids are fundamentally built from diamond, these systems have remarkable interactions with electrons. This new project was jump-started right away, when, in collaboration with Z.X. Shen in Applied Physics, Melosh found that a single monolayer of diamondoid molecules could cause a metal surface under high-energy photon bombardment—which normally emits electrons with a broad distribution of range of energies—to emit a nearly monochromatic energy beam. That such a thin layer of molecules (less than 1 nm tall!) could cause electrons to all dissipate their excess energy to relax to the diamondoid conduction band minimum was shocking and highlighted the uniquely strong interaction between electrons and diamondoid molecules. This discovery led to a Department of Energy funded diamondoid program that lasted more than a decade.

This research path led to a series of new discoveries about materials that had the size of

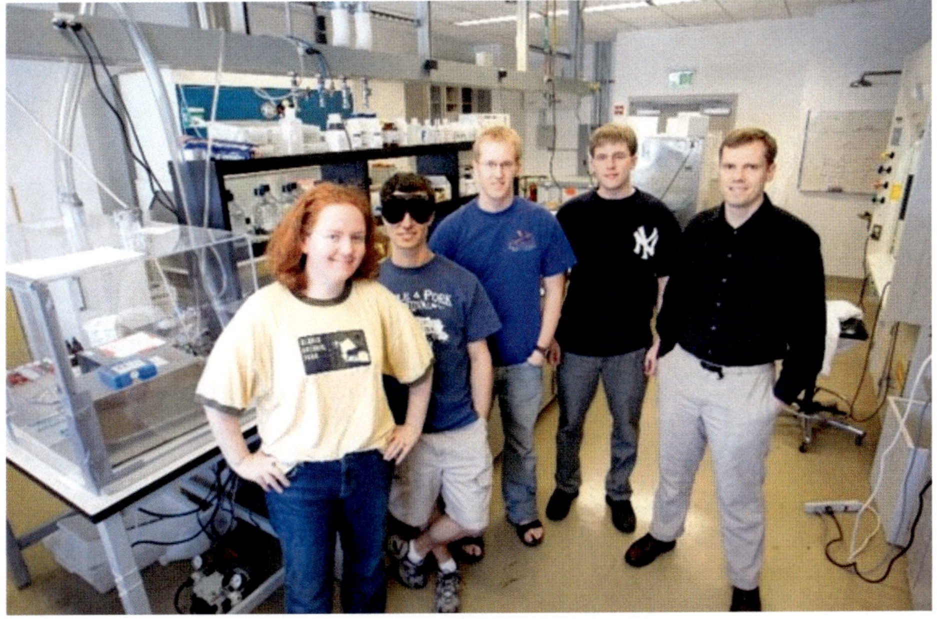

Nicholas Melosh, with his students in his lab, 2006. (School of Engineering)

small molecules, but the electronic and mechanical properties of nanoparticles. Diamondoids were found to provide the lowest-ever-reported work function for any air-stable coatings or organic molecule, and could field emit electrons at orders of magnitude lower voltages than typically used for conventional electron guns. Their remarkable electron emission properties led to a new idea in energy conversion, combining both photovoltaic energy and electron emission. This process, dubbed photon-enhanced thermionic emission, caught the attention of scientists and renewable energy companies worldwide. By capturing both thermal and photon energy, it combined the two major energy sources for electron generation in a way that had not been possible before. This led to a series of follow-up studies, supported by the Moore Foundation and ARPA-E, which looked at the fundamental physics and possible efficiency for these types of devices. The follow-up studies, in turn, led to the spin-off of Spark Thermionics, a thermal energy conversion start-up based in Berkeley.

Around 2010, Melosh's group realized that these diamondoids also should have remarkable self-assembly properties, a twist that completed an intellectual loop through electron transfer and back to molecular interactions and to the part of his group focusing on bio-interfaces. Since the diamondoids have a large number of atoms per molecule and very little entropy, their effective van der Waals interactions are incredibly strong. In addition to rapid crystallization, this attraction between diamondoids can be used to force other species together, with molecular-level control. The Melosh group successfully demonstrated synthesis of the smallest possible solid inorganic nanowires, with just a three-atom cross-section, yet millimeter-long. These are now being explored as designer one-dimensional materials, allowing combinations of materials and bonding configurations that are not possible through other means.

Most recently, diamondoids have been a central player in exploring new mechanico-chemical reactions and the nature of nucleation and growth. Combining atoms to make new molecules and materials by simply pressing them together, much like Lego blocks, has long been a dream for nanotechnology. However, it's never really been feasible, in large part because holding the atoms at the right positions and applying force along the correct directions has not been possible. However, the Melosh group developed an approach using the diamondoids to act like "molecular anvils," locally compressing molecular systems at the atomic scale, and directing chemical reactions. Here the diamondoids essentially convert an isotropic stress into an anisotropic strain at the molecular level, driving the atoms to come together in particular configurations. While not quite as versatile yet as a 'nano-assembler', this opens the door for precision molecular synthesis and mechanically driven reactions, such as more efficient CO_2 reduction.

Alongside all the diamondoid research has been an equally vibrant exploration of biologically related self-assembly at organic-inorganic interfaces. The core idea has been to investigate how nanoscale inorganic devices, such as those fabricated through CMOS processes, could intimately and seamlessly integrate with biological cells and tissues. A large part of this work has focused on understanding how to get through the outer barrier of the cell, the lipid membrane. For example, the team looked at how metallic electrodes decorated with a super thin, 5-nm tall band of hydrophobic species would insert into a lipid bilayer of a cell. These 'stealth electrodes' could covertly lodge within the interior of the cell membrane, without harming the cell. The first example of a large inorganic material acting like an artificial membrane protein, allowing passage of electric fields across the membrane, this effort has led to demonstrations of the technology for use in cell analysis and drug screening.

While impressive, the stealth electrodes only allowed passage of electrical currents, yet the natural language of cells is chemical. In 2011, Melosh's team first introduced 'nanostraw' arrays, consisting of 100-200 nm diameter 'straws' protruding from a plastic sheet that could pierce the cell wall without perturbing the cell. These were built using a recently introduced technique in the CMOS industry, atomic layer deposition, allowing conformal deposition of a very thin layer of glass from the gas phase. Together with standard water filtration membranes, the process produces large areas of nanoscale cell interfaces quickly and easily. Liquids, proteins, and genetic material placed below the plastic sheet can diffuse up through the nanostraws and into the cells. This was a breakthrough for cell interfaces, providing a rapid and simple means of getting materials into cells. This concept has been expanded upon in subsequent years, showing high efficiency cell transfection, non-natural cargo delivery, and electrically assisted injection of large pieces of DNA, and has been spun out as Navan Technologies, Inc, a start-up company for cell transformation.

The next breakthrough for the nanostraws came in 2017, when the Melosh group demonstrated that not only can you deliver things into a cell, but you can also extract materials out of a cell. This is akin to a blood draw, but at the single cell level, taking a tiny sample out for analysis yet not killing the cell in the process. This first-of-its-kind demonstration provides a new window into the operation and activities within normally behaving cells, and the group's announcement of the process was awarded the National Academy's Cozzarelli Prize for best paper in 2017.

Most recently, the Melosh group has moved into brain-machine interfaces. These are systems to help treat neurodegenerative diseases, such as Parkinson's or ALS, or to help amputees and paralysis victims better control neuro-prosthetic devices. Current devices are overly large and stiff, eventually causing tissue damage and loss of signal. Using modern semiconductor fabrication techniques, the Melosh team is taking new approaches to develop devices with lower mechanical stiffness, yet vastly larger data throughput. The team hopes to create new technologies for fundamental understanding of brain circuits, leading into effective new visual and auditory prostheses.

Alberto Salleo ~ Using Unconventional Semiconductors for Thin Film Electronics

Alberto Salleo joined the Materials Science Department on December 1, 2005. He spent his first day on the job at the MRS Fall meeting in Boston, keeping an eye out for potential future post-docs. Alberto had worked at PARC, Inc. (formerly known as Xerox PARC) since 2001, first as a post-doc and then as research staff member, on the use of molecular materials as semiconductors for thin film transistors. At Stanford, he was the last addition to a strong team of young faculty working on the same topic, from synthesis (Zhenan Bao) to device physics (Michael McGehee) and engineering (Peter Peumans). This strongly synergistic environment helped him grow his group in different directions and he soon found himself interested in studying fundamental structure-property relationships in these materials.

Alberto Salleo (A. Salleo)

Organic or molecular semiconductors are different from the more conventional inorganic ones, such as silicon (Si) or gallium arsenide (GaAs) inasmuch as they are kept together by weak van der Waals interactions. This weakness is turned into a strength because these materials can be processed from liquids at low cost and low temperature. Furthermore, they are soft and compliant, ideally suited for flexible and stretchable electronic devices that one can wear on the skin—or that can even mimic skin itself. While scientists and engineers have been able to synthesize new materials with interesting performance metrics, a theory of how charges and electronic excitations move through these materials does not exist. Indeed, they are the typical in-between case, being neither perfectly crystalline nor amorphous, which makes building a generalized theory very difficult. In a community dominated by chemists and physicists, Salleo found a niche as a "card-carrying" materials scientist by applying advanced characterization techniques to understanding how structural defects affect the materials' properties.

At Stanford. Salleo was fortunate to obtain extended access to the Stanford Synchrotron Radiation Lightsource (SSRL) at SLAC and could collaborate extensively with a world-renowned expert on x-ray scattering in soft matter, Michael Toney. As a result, within a few years Salleo's group gained an international reputation for being able to perform the most in-depth x-ray analysis of semiconducting organic polymers and, importantly, connect these results to the physical properties of interest in devices such as transistors or solar cells. This aspect of his group's body of work is summarized in a 2013 paper, "A Universal Relationship Between Disorder, Aggregation and Charge Transport in Conjugated Polymers," where a lateral study across many materials is used to derive very general rules as to what is needed for effective charge transport in semiconducting polymers.

This work was further enhanced by adding a theoretical foundation, thanks to a very fruitful collaboration with Andrew Spakowitz, then in Chemical Engineering (now with a 50 percent appointment in the MSE Department). While photovoltaics was never a main focus of Salleo's research, the research environment at Stanford with its strong emphasis on energy research pushed him in this direction after a few years. Salleo built a very productive working relationship (and a personal friendship) with Michael McGehee. Working together, their two groups surprised the field by discovering an unexpected reaction intermediate in the charge generation process in organic photovoltaics. The groups found that the formation of an energetically relaxed intermolecular charge transfer state was a necessary precursor to the generation of free charges, a result in contrast to conventional wisdom at the time, where only highly excited charge transfer states were thought to be able to lead to photocurrent. This result led to a better understanding of the device physics of organic photovoltaics.

In the last few years, Salleo's research has branched out towards applying organic semiconductors in bioelectronics. Among the most notable recent results are the demonstration of a wearable patch that measures cortisol, known as the "stress hormone," in human sweat; and the development of a completely new electronic device that mimics the functionality of synapses and can be used for brain-like computing.

Salleo's training in thin film electronics was acquired during his post-doctoral position at PARC. While a graduate student in Materials Science and Engineering at UC Berkeley, with Timothy Sands as his doctoral advisor, he spent time doing research at Lawrence Livermore National Laboratory (LLNL). At LLNL, Salleo studied the origins of optical breakdown in synthetic silica glass for the National Ignition Facility (NIF). NIF is a "Big Science" project aiming to achieve nuclear fusion by shooting 192 laser beams at a small target containing a mixture of frozen deuterium and tritium. In addition to the enormous scientific challenges that involve generating enough pressure to force the nuclei to fuse, NIF faced a tremendous engineering challenge that became its Achille's Heel. The lenses used to focus the laser beams were pitting and eventually cracked when exposed to the high-power IR and UV laser beams.

Just as in most materials science problems, it turned out that the optical weakness of the lenses was due to defects. With his research advisor, Salleo showed that sub-surface cracks created during the harsh polishing process could concentrate light up to a factor of 20 (or even 100 for conical cracks), thus locally exceeding the intrinsic optical strength of the ultrapure glass used to manufacture the lenses. He also found that during the optical breakdown process the local temperature and pressure can be so high as to transform the amorphous silica into a high-pressure crystalline polymorph, stishovite, usually found in meteorite-impact craters. It was particularly gratifying to him to see, many years later, that a colleague in MSE at Stanford, Evan Reed, performed calculations that confirmed

Alberto Salleo, with graduate student Scott Keene, characterizing the electrochemical properties of an artificial synapse for neural networks computing. (School of Engineering)

the occurrence of the transformation and explained its mechanism at the atomic level.

After working on laser-materials interactions, Salleo decided that he wanted to learn something completely new, and got a post-doc offer to join Robert Street's group at PARC to be part of their first team dedicated to working on printed electronics. As luck would have it, Street had worked on glasses for a long time through his interest in amorphous semiconductors. This is probably why he hired Salleo in the first place, even though Salleo says that he did not even know how a transistor worked when he started as a post-doc. The newly formed team at PARC was very dynamic, and at the time printed electronics was a fast-growing area. While a post-doc, Salleo published his research on electrical instability in organic semiconductors and charge trapping.

The environment at PARC was extremely stimulating as Street was adamant about pushing fundamental science and technology in parallel. As a result, the team at PARC was conducting low temperature charge transport experiments in a cryostat in one room while also, in the next room, building an all-printed transistor array to be used as a display backplane. The experience at PARC was very useful for Salleo as it really brought home the point that technological development is part of an ecosystem that must contain fundamental science to be robust. Because he was able to publish in peer-reviewed journals and speak at conferences, Salleo was ready to seize the opportunity when a faculty opening occurred at Stanford. He was very excited by this opportunity, as he had already struck up friendships with several MSE faculty members (notably Mark Brongersma, Michael McGehee and Nicholas Melosh), and knew how supportive the environment for assistant professors was at Stanford. Salleo never regretted delivering the faculty application materials to Billie Kader by hand at 4pm on the day the search closed.

Salleo was born in 1971 in Prague, then Czechoslovakia. His father is Italian (Sicilian) and his mother is Hungarian. Alberto's mother had escaped Hungary in 1947 and grew up in Colombia. According to his mother, as a friendly gesture to a foreigner, Alberto was born in a broom closet outfitted with a bed, rather than in one of the communal birthing rooms then *en vogue* in communist countries, As the son of an Italian diplomat (his father was the Italian Ambassador to the US from 1996 until 2003), young Salleo lived in Czechoslovakia (1971-72), the US (1972-74), Italy (1974-77, 1982-85, 1988-95), Germany (1977-81), and France (1986-88). While in college, he visited the USSR and later, Russia, when his father was posted to Moscow. He was in Red Square when the Soviet flag was taken down and replaced with the Russian flag (his memory of what happened later is a little blurry due to vodka-fueled celebrations with locals).

Salleo was schooled in the French system because it is completely standardized across the world, making it ideal in situations where one might move in the middle of the academic year. The French system is very rigorous, even rigid, and places much emphasis on math and science, using math as a tool to rank students regardless of their ultimate intellectual inclinations.

Towards the end of his junior year, Salleo knew that he would study something scientific in college. After graduating with the French baccalaureate, he faced the choice of either entering the punishing French preparatory school system in order to apply to elite French engineering schools or attending the more relaxed public university in Rome. He opted for the latter, but then hesitated between majoring in Physics or Chemistry before ultimately choosing Chemistry. In this respect, Materials Science best represents his intellectual interests, as it perfectly conjugates these two disciplines while adding an applied flavor to them.

Salleo enrolled at the University of Rome, *La Sapienza,* at the time the third largest in the world, with 180,000 students attending. The

Chemistry Department had approximately 500 first-year students, and the five-year program culminated with a one-year long thesis, making the curriculum more like that of a master's degree than a bachelor's degree. He chose a physical/theoretical chemistry concentration, mostly due to his aversion for instructional labs. He is still fondly remembered by his classmates for causing a small explosion that, though otherwise harmless, ruined all experiments in a 20-foot radius of the sealed container he forgot he had left on an active hot plate. It is thus only fitting that at Stanford he teaches an undergraduate/master's instructional lab on thin film electronic devices. Interestingly, his is the only lab class at Stanford where undergraduates can make electronic devices. In the heart of the Silicon Valley, Stanford undergraduates make transistors using polymers, and learn that silicon is a very convenient flat substrate for this purpose.

For his thesis, Salleo joined the Center for High–Temperature Thermodynamics at the University of Rome to study high-temperature oxidation of refractory carbides. This started his lifelong love for classical thermodynamics, which led him to take on the mandatory graduate course MSE 204, "Thermodynamics and Phase Equilibria" at Stanford. Salleo has worked hard at perfecting this class, climbing from a ranking in the student surveys that put him in the 7th percentile all the way to winning the 2016 Gores Award for Excellence in Teaching. In his last six months of college, Salleo accepted an offer to join L'École Polytechnique, outside Paris, as a visiting student. As he took classes in solid-state physics and materials chemistry, while doing research in the famed metallurgy lab at Orsay where Friedel had worked, Salleo realized that he was really interested in materials and defects.

When he returned to Rome, he applied for a Fulbright Fellowship, which he won. With his fellowship in hand, he decided he would apply to Materials Science Departments rather than Chemistry Departments. He was accepted at UC Berkeley, and joined the graduate program there in the Fall semester of 1995. In addition to the research, his most memorable experiences included Bill Morris' graduate thermodynamics lectures and TA-ing the undergraduate thermodynamics class for Andy Glaeser. The perfect combination of physics, chemistry and real-world materials applications of Berkeley's materials science curriculum cemented in him a lasting passion for this discipline. His enthusiasm for interacting with students, through lectures or research, and his love of materials science are perfectly embodied by Sir Charles Frank's famous quote "Crystals are like people, it's defects in them that make them interesting."

In 2019, Salleo was appointed the 11th chair of the Department of Materials Science and Engineering.

Chapter 23

Appointments for Energy and the Environment

By the 21st century, the department is no longer making faculty appointments to replace departing faculty or to fill either specific teaching needs or perceived gaps in the coverage of materials science fundamentals, as had been done in the past. Instead, although target materials science subjects are mentioned in the faculty search announcements, appointments typically go to those who appear to be doing the most exciting science overall. Thus, individuals are selected based more on the likely impact they would have on important societal problems in which materials play a role, and less on their demonstrated ability to advance the core discipline of materials science. With this objective in mind, several appointments have been made to respond to perceived needs in the production, storage and utilization of energy and related environmental needs.

Yi Cui ~ Nano for Energy and Environment

Yi Cui joined the Materials Science faculty in 2005 as an assistant professor, after finishing his Miller Postdoctoral Fellowship at the University of California, Berkeley. He works in the area of designing nanomaterials to address energy and environment problems.

Yi Cui was born in the small city of Laibin, in Guangxi Province of China, on January 22, 1976. In China during the Cultural Revolution (1966-1976) many educated people, including his parents, were relocated from urban centers to rural villages. Cui spent much of his first five years in a village, and then moved back to Laibin city with his parents when they were released from the village. He started his elementary school education soon after. His father was a high school chemistry teacher and his mother was an elementary school Chinese teacher. Despite the very poor economic conditions of the time, his education was not neglected. Influenced by his father, he developed a strong interest in science, particularly in chemistry. He was ranked first in his class throughout middle and high schools.

In 1993, Cui was admitted to China's University of Science and Technology (USTC) for undergraduate study, majoring in chemistry. USTC is one of the top universities in China, recruiting the brightest high school kids from across the country. At the time, it had a rigorous five-year undergraduate program in which all students, no matter their major, were required to take a large number of credits in mathematics and physics, giving USTC a well-earned

Yi Cui during his time at Harvard, 1999. (Yi Cui)

reputation for graduates possessing solid fundamentals.

Most of the USTC undergraduates go on to graduate school in the United States. In 1998, Cui went to the Department of Chemistry and Chemical Biology at Harvard University for his Ph.D study in Physical Chemistry. He joined the group of Professor Charles M. Lieber, a world-leading figure in nanotechnology. Cui was a pioneering Ph.D student there, as he started a new research field with Lieber on nanowires. His two publications have helped to define the field of nanowires and are frequently cited: "Functional Nanoscale Electronic Devices Assembled using Silicon Nanowire Building Blocks," in Science (2001), and "Nanowire Nanosensors for Highly-Sensitive, Selective and Integrated Detection of Biological and Chemical Species," *Science* (2001). He was awarded the Gold Medal Graduate Student Award in Materials Research Society in 2001.

After completing his Ph.D., in 2002 Cui was offered the prestigious Miller Fellowship to become a post-doctoral fellow at the University of California, Berkeley. While there, he worked in the lab of Professor Paul Alivisatos, another leading figure in nanotechnology. Alivisatos is credited with co-starting the field of colloidal nanocrystals. At Cal, Cui developed capillary force assisted assembly of nanocrystals and measured the low temperature transport of nano-tetrapods, which show interesting electronic coupling behaviors.

In 2004, Cui, then 28, was awarded the World Top Young Innovators Award by *Technology Review*. During his postdoctoral work, he generated numerous research ideas he has pursued in his independent research career, including an idea of using electrochemical measurements at nanoelectrodes to determine single biomolecule configurations and dynamics, and to study phase changes at single nanostructure levels. He received many faculty position offers, but decided to join Stanford's Materials Science Department, he says, because Stanford offered a perfect combination of a top research program, a collegial environment, a tradition of entrepreneurship, and very nice weather. Cui arrived at Stanford on September 1, 2005.

Cui had expected to pursue the nanoscience agenda he had followed for much of his time as a post-doc. His plans changed significantly, however, during his last year at Berkeley when Stanford professor of physics Steven Chu moved to Berkeley to become director of Lawrence Berkeley National Laboratory. There, Chu launched multiple major initiatives centered around clean energy. Already, the issue of CO_2 greenhouse gas emissions was recognized as an urgent problem for earth sustainability, and Chu was encouraging scientists, especially young ones, to work on energy-related problems. Influenced by Chu, Cui decided to dedicate his research effort toward clean energy when he moved to Stanford.

At Stanford, Cui quickly started what would become a world-class research program on nanomaterials for energy. The most important breakthrough came in a 2008 paper published in *Nature Nanotechnology*, "High-performance lithium battery anodes using silicon nanowires." This pioneering work triggered a worldwide explosion of research using nanoscience for energy storage. His work brought a $10 million research grant to serve as a King Abdullah University of Science and Technology (KAUST) Investigator for five years. His Stanford group also grew quickly, becoming the largest in the department, with 40 to 50 graduate students and postdocs. Despite a heavy research load, it is not uncommon for Cui to stroll into the office of another faculty member, with nothing in his hands, to chat briefly and pass on some words of encouragement.

Cui has made seminal contributions on materials design for high energy density batteries, grid-scale storage and battery safety, and has worked collaboratively with Robert Huggins, an expert in solid state ionics. It has turned out that the new generation of battery materials has opened up exciting scientific opportunities in mechanics due to their large volume change. Cui formed a close collaboration with the author of this piece, generating important insights on the electrochemical and mechanical coupling in energy storage materials. He also worked closely with Zhenan Bao in Chemical Engineering, bringing new concepts of soft materials, such as the self-healing polymer, into the battery field.

In addition to energy storage, Cui has built up very successful research programs in such diverse topics as solar cells, transparent conducting electrodes, two-dimensional layer materials, electrocatalysts, in-operando electron microscopy, cooling and warming textiles and nanobiology. He has also initiated new ideas on nanomaterials for air filtration, water filtration and soil cleanup. Most recently, he pioneered the idea of using state-of-art cryogenic electron microscopy for materials research, which led to a new understanding of fragile materials. This large intellectual footprint has opened up a wide range of active collaborations across the Stanford campus, particularly with Steven Chu in Physics, Shanhui Fan in Electrical Engineering, Craig Criddle and Alexandria Boehm in Environmental Engineering, Mark Brongersma and Michael McGehee in Materials Science, and Michael Toney at SLAC

His scientific leadership in materials for energy and environment prepared him to play a leadership role at major research centers. In 2011, he served as a co-director of the Bay Area Photovoltaic Consortium, where he brought in $25 million in research funding from the U.S. Department of Energy (DOE), with additional support from industry and universities. The project was part of the DOE SunShot Photovoltaic Manufacturing Initiative (PVMI). In 2016, he served as a co-director of Battery500 Consortium, bringing in $50 million from DOE. These major centers have served as vibrant research platforms to support many Stanford faculty as well as other faculty around the nation, to work on important energy problems.

Cui has published more than 420 research papers, filed more than 50 patent applications and has given some 400 plenary, keynote, and invited talks. Among the most highly cited

Yi Cui, 2011. (Stanford News Service)

scientists in the world, he was ranked no.1 in Materials Science in 2014 by Thomson Reuters in their list of "The World's Most Influential Scientific Minds." His three inventions (nanofilters for water disinfection, thermal batteries, cooling textile) have been named "Top 10 World Changing Technology" by *Scientific American* in 2010, 2014, and 2016, respectively. He is an associate editor of *Nano Letters* and serves on the editorial boards of numerous scientific journals. He was elected as a Fellow of Royal Society of Chemistry (2015); a Fellow of Materials Research Society (2016); and a Fellow of Electrochemical Society (2018).

Among his numerous awards is his selection as Blavatnik National Laureate in Physical Sciences and Engineering (2017), a distinction given to a single individual each year in United States. Other major awards include the MRS Fred Kavli Distinguished Lectureship in Nanoscience (2015); the Small Young Innovator Award (2015); the Inorganic Chemistry Frontiers Award for a Young Scientist (2015); and the Inaugural Schlumberger Chemistry Lectureship (University of Cambridge, 2015).

During his 14 years at Stanford, Cui has trained some 100 Ph.D. and post-doctoral students in his group. Among them, about 70 have become faculty members in the United States and around the world, including: Liangbing Hu at the University of Maryland; Jie Yao at the University of California Berkeley; Yuan Yang at Columbia University; Matthew McDowell at Georgia Institute of Technology; and Guihua Yu at the University of Texas, Austin. His students have won 15 Materials Research Society Graduate Student Awards; two Materials Research Society Postdoctoral Awards; and six TR35 World Top Young Innovators Awards.

Cui has founded three companies. He founded Amprius Inc. in 2008 to commercialize his silicon (Si) anode high-energy battery technology. The high-energy batteries invented by him are now used in commercial products, which could potentially revolutionize portable electronics, drones, and electric transportation applications. In 2015, he and Steven Chu co-founded 4C Air Inc., to commercialize their breakthrough technology to remove particle pollutants from the air. In 2017, he co-founded EEnovate Technology Inc. to serve as a technology incubator to commercialize his group's developments in water, textile, large-scale energy storage, and new materials technology.

William Chueh ~ Materials Science for Energy Production, Storage and Utilization

William Chueh was the first faculty member to be jointly appointed between the Precourt Institute for Energy and an academic department. This bridge provided him a fertile ground for interdisciplinary research rooted in materials science aimed at addressing the most pressing societal challenges related to energy production, storage, and utilization. Chueh was trained in point defects in solids, thermo-chemistry, and electrochemistry. When he began his career at Stanford in 2012, Chueh had proposed to work in the field in solar fuels (converting sunlight, water, and CO_2 to chemical fuels). Inspired by colleagues working on both fundamentals and applications of energy materials, he rapidly expanded his research activities to include energy storage, particularly batteries, and electrocatalysis. In 2018, he was recognized by the Materials Research Society Outstanding Young Investigator Award, the third in the department to receive the prize, for his "groundbreaking research on ionic and electronic charge transport and interface chemistry relevant to electrochemical devices."

William Chueh, 2012. (MSE Dept.)

Chueh is a practitioner of the theory-synthesis-characterization loop central to materials science and carried out fundamental studies on redox reactions involving protons, alkali metal ions, and oxygen ions. He is particularly passionate about developing characterization tools that embrace the dynamical and heterogenous nature of energy materials. For example, he has developed in-situ x-ray microscopy for tracking battery charging and discharging at nanometer resolution, which reveal phase separation and diffusion pathways.

In another example, working with Michael Toney at SLAC, he has combined x-ray diffraction, spectroscopy, and scattering to understand point defects in layered oxides, which are used as positive electrodes in lithium-ion batteries. Early in his career, Chueh was often seen late at night with his students at the Stanford Synchrotron Radiation Lightsource (SSRL) and the Berkeley Advanced Light Source (ALS). He is an advocate for large-scale science facilities; in 2017, he briefed the United States Congress on the importance of synchrotron radiation facilities to the national interest.

Although Chueh's research is mainly fundamental, he has been strongly influenced by the entrepreneurial spirit of Stanford students. Working closely with industry, Chueh and his students have appreciated the challenges associated with turning laboratory innovations into commercial realities. Ins 2016, he received the Science Award Electrochemistry from BASF and Volkswagen for his contribution towards understanding redox processes in lithium-ion batteries, particularly in connection to electric vehicles. Beyond electrifying transportation, Chueh has also recognized the immense challenges associated decarbonizing the electrical grid. To do so, storing renewable electricity from solar and wind in a cost-efficient manner is paramount. With his student Antonio Baclig, he developed a room-temperature liquid metal flow battery that promises to deliver long lifetime, high energy density, and low cost. Chueh's desire to study fundamentals of redox process and applying insights to solve pressing problems in energy and climate has been heavily influenced by his faculty mentor Michael McGehee, who shared the same passion.

On the teaching front, Chueh likewise highlights the connection between fundamentals and application. Undergraduate thermodynamics is a challenging topic to teach in any engineering department, because the concepts are often abstract (*e.g.*, entropy) and the examples dated (*e.g.*, a refrigerator). When he took over as the instructor for MATSCI 154, Chueh broke from the usual mold of starting with the First and the Second Laws, etc. Instead, he divided the course into four modules, with each one focusing on an energy conversion technology: CO_2 capture; fuel cells; solar thermal electricity; and lithium-ion batteries. The lectures present the basic concepts that underlie the technology, and rigorously derive the thermodynamic limits. Using CO_2 capture as a vehicle to teach entropy of mixing, and using batteries to teach phase transformation, takes full advantage of the students' interest in these technologies. Fittingly, the class is titled "Thermodynamic Evaluation of Green Energy Technologies." Chueh has involved his graduate students in teaching. Yiyang Li, who contributed to developing the thermodynamics course, received Stanford's highest teaching honor, the Walter Gores Award (which is awarded to both faculty and graduate students).

By 2018, Chueh had graduated his 12th PhD student, three of whom received the MRS Gold Medal Graduate Student Award: Yiyang Li (2015), Xiaofei Ye (2015), and William Gent (2018). Chueh himself has received numerous awards, including Camille Dreyfus Teacher-Scholar Award (2016); Sloan Research Fellowship (2016); NSF CAREER Award (2015); and Solid-State Ionics Young Scientist Award (2013). In 2012, he was named as one of the "Top 35 Innovators Under the Age of 35" by MIT's *Technology Review*. William C. Chueh was born in Taipei, Taiwan, on December 22, 1982. He immigrated to Los Angeles at the age of 10, speaking only a few words of English. His first stint in research was during his junior year at Troy High School (Fullerton, California), when he spent two months working as a summer intern at the plasma laboratory at University of California, San Diego, and the Tokamak at General Atomic. He received his B.S. in applied physics, M.S. and Ph.D. in materials science with Sossina Haile, all from Caltech. Shortly before receiving his doctorate, in 2010, Chueh accepted an offer to join the MSE department as a faculty member. Before formally starting his position, in 2012, he spent two years at Sandia National Laboratories in Livermore, California, as a Truman Postdoctoral Fellow, training with Kevin McCarty, Tony McDaniel, and Farid El Gabaly.

Chapter 24

Appointments in Biomaterials

Although courses in polymeric materials and biomaterials had been offered in the past by Donald Lyman, who served as a lecturer on loan from the Stanford Research Institute in the 1960s, and by Richard Wallace, who served as a visiting assistant professor for a few years in the early 1970s, no serious commitment to soft materials was made until Sarah Heilshorn's 2006 appointment to the faculty. With the growing recognition that soft biomaterials were becoming one of the hot topics in materials science, two more appointments were made in this general area: Eric Appel and Guosong Hong. In addition, more and more faculty members began turning at least some of their attention to materials science problems in biomaterials. As the huge fields of biology and medicine attract the attention of MSE faculty, it seems likely that these trends will continue unabated for years to come.

Sarah C. Heilshorn ~ Biomaterials Pioneer and First Woman on MSE Faculty

Sarah Heilshorn was appointed to the MSE faculty in the Fall of 2006 to begin building her independent laboratory at Stanford. Her group designs biomimetic materials that are used both to study fundamental phenomena involving self-assembly and to develop novel medical therapies. When she first joined the department, hers was the only lab with a focused interest in biomaterials. Indeed, at that time most Materials Science & Engineering Departments throughout the United States had not yet invested in biomaterials research. Thus, the department and the university made a substantial commitment to build a new tissue engineering and protein engineering facility for Heilshorn when she arrived.

In addition to requiring all types of new facilities, Heilshorn also needed to recruit new types of students. Since the department had not yet attracted students with an interest in biomaterials, her first Ph.D. advisees were from the Departments of Chemical Engineering; Mechanical Engineering; and Chemistry; and from the newly created Bioengineering Department. Her scope of research is quite interdisciplinary, and she currently continues to advise students from these departments, along with students from MSE; the Stem Cell Biology Program; and the M.D./Ph.D. Program.

Her research has focused on the design of new materials to interface with stem cells, a specialized cell type that can give rise to many different types of cells. To achieve this, she often creates her materials to mimic key aspects of the proteins that have naturally evolved to interface with stem cells. Key contributions from her group include the design of injectable gels for regenerative medicine, the development of materials to aid in the manufacturing of large numbers of stem cells, and the fabrication of scaffolds for tissue engineering and 3D bio-printing. For this creative work, she was recognized with a National Science Foundation Career Award and the National Institutes of Health New Innovator Award. She was also elected as a Fellow of the American Institute for Medical and Biological Engineering, a Fellow of the Royal Society of Chemistry, and to the Board of Directors of the Materials Research Society.

Heilshorn was born in 1975 in the rural community of Defiance, Ohio, the same small city where her parents were born and raised, and a city where everyone seemed to know everyone else. Her father was a metallurgist in the local General Motors foundry and a football coach at Tinora High School, where Sarah attended. Her mother was a homemaker during Sarah's early years, and then a college student and an office manager. Growing up, Sarah enjoyed school, along with a wide variety of extracurricular activities including dance (ballet, tap, jazz, and modern), athletics (basketball and cross-country running), music (singing, playing piano, and playing percussion in the marching band and orchestra), and visual arts. Her parents instilled in her a love for travel and for learning about different places and people. Sarah had the chance to take her first international trip as a junior in high school when she received a Congress-Bundestag Exchange Student Fellowship, allowing her to spend the next year studying and living with a host family in Rheinbach, Germany. She graduated from Tinora High School in 1993.

At the time, she was unsure of what field of study she should pursue in college. Since she had always enjoyed her math and science classes, her parents encouraged her to consider engineering. They considered this a much more secure choice than studying foreign languages or the arts, which she also greatly enjoyed. Following their advice, after earning a President's Scholarship, she enrolled in the fall at Georgia Tech in Atlanta as a chemical engineering major. At Georgia Tech, Heilshorn was a co-op student, and beginning the summer after her freshman year, she was placed as an intern with a research laboratory at Procter & Gamble in Cincinnati, Ohio. For the next three years, she would alternate spending one quarter in Atlanta on coursework and then one quarter in Cincinnati working at Procter & Gamble. This arrangement allowed her to supplement her scholarship with money to pay for her books and living expenses on campus. She greatly enjoyed the research environment at Procter & Gamble, which was her first exposure to creative, scientific problem solving.

During her quarters on campus, Heilshorn joined Phi Mu Sorority, Omega Chi Epsilon Honor Society, and was selected to be a Georgia Tech Ambassador. She enjoyed a diverse array of college life activities including giving campus tours, organizing events at the Student Center, participating in Homecoming traditions such as the Wreck Parade and the Mini 500 Tricycle Race, and volunteering at a local elementary school and a women's shelter. After completing her contract with Procter & Gamble, she did further research internships at Texaco, Exxon, and at Georgia Tech with Professor Jeffrey Morris in his complex fluids research group. During most of her time as an undergraduate student, Heilshorn never thought of continuing her studies into graduate school. However, Morris encouraged her to consider applying to doctoral programs. She graduated in 1998 with a degree in chemical engineering and certificates in German and International Affairs. On a whim, she successfully applied to Caltech, since it was the only school that didn't have an application fee. She started her doctoral studies at Caltech the following fall as a National Science Foundation Graduate Fellow.

Heilshorn was planning to study environmental chemistry at Caltech. Her stints at Texaco and Exxon had left her with the feeling that more scientists should be working to preserve our environment. However, after hearing a seminar by Professor David Tirrell, who had just moved his laboratory from the

Sarah Heilshorn, 2014. (School of Engineering)

University of Massachusetts, Amherst to Caltech, she became interested in biomaterials. Tirrell was developing new methods to create precise polymers using protein engineering technology. Heilshorn was fascinated, and she decided to try to switch her research interest. She was thrilled to be offered a position in Tirrell's research group and set to work teaching herself biology—a subject she had previously only studied once—as a freshman in high school. Her thesis work was on the study and design of elastin-like proteins, which were being investigated as a medical implant material.

While a student at Caltech, Heilshorn also continued her interests in the performing arts by joining the Women's Glee Club chorus, helping to start the Caltech Dance Troupe, and taking dance classes at the University of Southern California. It was also at Caltech that she met her future husband, Andrew Spakowitz, a fellow Ph.D. student in the chemical engineering department who would also go on to become a faculty member at Stanford. One day, as Heilshorn and a group of other women Ph.D. students were chatting about how little they knew about social science data that backed up their own personal experiences about the challenges of being a woman in science, they decided to start a new student group, Women in Engineering, Science and Technology (WEST). The WEST group organized a reading club to discuss the scientific literature on women's studies. They also sponsored a broad range of social activities for women students and faculty to meet one another. It was during these meetings that Heilshorn first began to consider a career in academia.

During her graduate studies, Heilshorn applied to the National Science Foundation East Asia Summer Internship Program. With her advisor's help, she succeeded in being placed in an internship with Professor Tetsuji Yamaoka, then at the Kyoto Institute of Technology. That summer, she continued her studies of elastin-like proteins and was able to do a great bit of touring around Japan and to give several seminars at different Japanese universities. Upon returning to Caltech, she was selected for the campus-wide Everhart Lectureship. At this time, she began to seriously consider applying for faculty positions. In 2004, she interviewed and was selected for an assistant professor position in the Materials Science & Engineering Department at Stanford. She graduated from Caltech with her doctorate that spring, and delayed the start of her faculty position for two years so that she could gain extra training as a postdoctoral scholar at the University of California, Berkeley.

In 2005, Sarah Heilshorn and Andrew Spakowitz were married in Grand Rapids, Ohio. They were both pursuing postdoctoral studies in Berkeley while preparing to start their independent labs at Stanford. Heilshorn worked with Professor Mu-ming Poo in the Molecular and Cell Biology Department. There she was introduced to a range of topics in neurobiology and worked on a project to identify the mechanisms that neurons use to organize themselves into functional circuits. During this time, she also taught herself how to design and fabricate simple microfluidic devices, which led to the funding of her first research proposal by the National Institutes of Health.

She began her appointment at Stanford in fall 2006, when she taught a new version of the undergraduate course "Introduction to Materials Science" by introducing each fundamental topic with a case study from medical materials. Although her research facilities in the Moore Building were not yet ready, she also began advising research trainees right away by using temporary space in the Mechanical Engineering Research Lab.

Over the next eight years, the Heilshorn lab continued to grow, in size and impact, while Heilshorn also added new courses to her teaching repertoire. She taught the core graduate course "Organic Materials," the elective graduate course "Biomaterials in Regenerative Medicine," and the undergraduate introductory seminar "Bioengineering Materials to Heal the Body."

During this time, the Heilshorn-Spakowitz family also grew. In July 2011, their first son, Cyrus Spakowitz, was born and adopted, followed by the adoption of a second newborn son, Oscar Heilshorn, in November 2012. Both boys attended preschool on campus, allowing the whole family to carpool together to Stanford. Heilshorn earned tenure in 2014, and in 2015 the family went on a sabbatical, first to the University of Chicago and then to the University of Sydney, New South Wales, Australia, where they had several great adventures.

Since returning to Stanford, she has taken part in many activities to recruit and support a diverse group of undergraduate, graduate, and postdoctoral trainees on campus. For her service to undergraduate education, including the remote teaching of *Introduction to Materials Science* to Stanford students studying abroad, Heilshorn was named a Bass University Fellow in Undergraduate Education. Through her mentorship in numerous different graduate and postdoctoral programs on campus, she is committed to creating a more diverse and inclusive educational environment at Stanford.

Eric A. Appel ~ Dynamic Interactions in Soft Materials

Eric Appel joined the MSE faculty in March 2016 to work at the interfaces between materials science, chemistry, and biology. The theme of his work has been to integrate concepts and approaches from these fields to develop new materials for addressing healthcare challenges of critical importance to society. Through his B.S., M.S., and Ph.D. training in the chemical sciences, his research has focused on the design, synthesis, and characterization of biomimetic polymeric materials. His later research has evolved to focus on exploiting the unique

Eric Appel, 2016. (MSE Dept.)

properties of these materials for biomedical applications, both to address long-standing challenges in the field and to enable new therapeutic opportunities.

Eric Appel was born in Long Beach, California, in 1984 to Stan and Cyndi Appel. His father was born in a small town in Kansas but had spent most of his life in Long Beach, while his mother was a California native, born in nearby Inglewood. Stan had received a B.S. in business management from California State University, Long Beach, in 1977 and worked for more than 40 years as a police officer and detective for the Long Beach Police Department and the Los Angeles County District Attorney's office. Cyndi was a home maker and worked part-time at a flower shop in Los Alamitos, California.

Appel attended public schools in Los Alamitos, California, and was fortunate to have numerous excellent teachers throughout his education, allowing him to develop his interests in science and math. He attended Cal Poly San Luis Obispo (SLO) to pursue chemistry, and later joined the BS+MS program in chemistry and polymer science. Since Cal Poly SLO is primarily an undergraduate institution, research projects were scarce, so he pursued his M.S. thesis research at the IBM Almaden Research Center, near San Jose, with Dr. James L. Hedrick and Dr. Robert D. Miller, on the development of new biodegradable polymeric materials for drug delivery applications.

During his time at IBM, Appel was exposed to the concepts of supramolecular chemistry, or "chemistry beyond the molecule," and he decided to pursue his Ph.D. research on polymer science and designed self-assembly. He went on to the University of Cambridge to work with Professor Oren A. Scherman; his thesis research focused on the synthesis and characterization of novel classes of dynamic polymer materials. At Cambridge, Appel was influenced by the breadth of top-quality research occurring across departments, and the ability to apply technologies with unique properties in diverse areas. As he says, the engineering requirements for a complex, stimuli-responsive hydrogel for use in hydraulic fracturing are very similar to those for an injectable hydrogel for sustained delivery of pharmaceuticals.

Following his Ph.D. in 2012, Appel was awarded a postdoctoral fellowship by the Wellcome Trust in the United Kingdom to work with Professor Bob Langer at MIT. While at MIT, Appel applied his training in design and synthesis to spearhead the development of a platform technology with broad interdisciplinary applications, such as implantable materials for controlled drug delivery and cell-based tissue engineering. Utilizing well-established nanoparticle technology originally developed in the Langer lab, and a simple chemical modification of a polymeric carrier, Appel developed a novel class of injectable hydrogels with unprecedented materials properties allowing for high drug-loading capacity and the potential for sustained release of diverse molecules.

Since the inauguration of his lab at Stanford, Appel has built a multi-disciplinary and collaborative research program aimed at generating new materials to address fundamental biological questions and engineer next-generation healthcare solutions. The Appel lab aims to incorporate fundamental topics in materials science, such as diffusion and mechanics, in the design of novel soft materials for a range of applications.

The breadth of Appel's research is illustrated by briefly considering some of his current projects. Therapeutic approaches that involve manipulation of the immune system to protect against infectious diseases or cancer have been successful, but they can be quite variable and can exhibit off-target toxicities. Improving vaccine technology to produce robust, targeted immune responses with specific immune cell types has emerged as a critical problem. There is much evidence for both prophylactic vaccines against infectious disease and therapeutic vaccines against cancer to suggest that combination immunotherapies using multiple compounds can lead to potent but safe treatments. Yet, these compounds must often be delivered alongside one another to appropriately coordinate the activation of immune cells. Critically, there is no single delivery platform able to achieve controlled co-delivery of a broad range of these molecules, The Appel group is developing a novel controlled delivery technology that is capable of precise and sustained delivery of diverse molecular reagents and that provides a basis for powerful, multifaceted immunotherapies.

Using some of the approaches developed in his biomedical research, Appel has developed a novel biomimetic hydrogel platform that can be used in numerous industrial applications, including as fire retardants. He has shown that these hydrogels can be tuned to persist in the environment for the duration of a fire season, and yet they are completely biodegradable during the off-season. Until now there have been no environmentally safe prophylactic fire-retarding materials available for use in fire prevention. Appel's materials can enable the prophylactic treatment of landscapes of particularly high fire danger, potentially preventing a vast number of wildland fires from starting in the first place.

Adhesions are fibrous bands of scar tissue that form between internal organs and tissues following any type of surgery. They arise from normal wound-healing processes but can lead to complications ranging from severe pain, infertility and impaired organ function. Current prevention technologies are extremely difficult to use and have not been widely adopted. Appel and his students are developing a way to prevent adhesion formation using a highly tunable viscoelastic hydrogel platform. By tuning the chemical and mechanical properties of these materials, they have developed a system that completely prevents adhesion formation in a large animal model of pericardial adhesions, where the current standard treatments completely fail.

Appel 's research experience had prepared him well for his first teaching assignment at Stanford: "Organic & Biological Materials" (MATSCI 210), which is a graduate level soft matter course. To expand the soft matter courses available to undergraduate students, Appel also developed a course entitled "Soft Matter in Biomedical Devices, Microelectronics, and Everyday Life" (MATSCI 158). This course fills a critical need within the curriculum across several departments at Stanford and is cross listed in the Departments of Chemical Engineering and Bioengineering. The goal of this course is to introduce students to the exciting developments in biomaterials and other soft materials research in a way that encourages them to pursue additional science and engineering training. The course emphasizes the role that soft matter can play in solving medical and other related societal challenges and provides an excellent vehicle for training the next generation of scientists and engineers.

Guosong Hong ~ Bridging Materials with the Mind

Guosong Hong joined the department in 2018 as an assistant professor, after doing post-doctoral work at Harvard. He had previously received his Ph.D. in Chemistry from Stanford. At Harvard he worked on a new form of nanoelectronics, called 'mesh electronics', with brain-tissue-like ultraflexibility that makes them fundamentally different from conventional rigid electronics and thus suitable for instrumenting the brain and other soft tissue. As a graduate student at Stanford, his research was focused on understanding and capitalizing on the exotic optical properties of carbon nanotubes, which absorb light in the visible range but emit fluorescence in another window of the electromagnetic spectrum. His developments made it possible to visualize deep brain structures in rodent animals. Thus, both his Ph.D. research at Stanford and his postdoctoral research at Harvard were focused on developing functional nanomaterials for interrogating and manipulating neural activity in live animals.

Hong is continuing that line of research as a faculty member at Stanford. His lab is developing functional biomaterials and optoelectronic tools to record and modulate neural activity in the brain, with the goals of bridging materials science and neuroscience, and blurring the distinction between the living and non-living systems. Hong will be teaching the course "Nanocharacterization Laboratory" (MSE 161), as well as a new course, "Materials Advances for Neurotechnology: Materials Meeting the Mind" (MSE 384).

Guosong Hong was born in Hefei, Anhui, China, on April 21, 1986. His father was employed by the Chinese Internal Revenue Service while his mother worked for a tobacco company. He received his elementary and secondary education in Hefei, graduating from Hefei No. 1 Senior High School in 2004. He enrolled that fall at Peking University in Beijing where he majored in chemistry. He received his B.S. degree in chemistry in 2008 and came to Stanford for graduate work in chemistry just a year later.

Guosong Hong (G. Hong)

As a graduate student at Stanford, Hong joined the research group of Professor Hongjie Dai in the Department of Chemistry. His research was on deep brain fluorescence imaging with low-bandgap optical materials, focusing on understanding and using the exotic optical properties of carbon nanotubes. These materials are intrinsically strong light absorbers in the visible range but emit fluorescence in another, previously unexplored, window of the electromagnetic spectrum, called NIR-II wavelengths. The peculiar optical properties of single-walled carbon nanotubes (SWCNTs) are associated with the various structures that these materials can have. In striking contrast to other fluorescent molecules, when SWCNTs are made water-soluble and biocompatible, they can be delivered into blood vessels for imaging at NIR-II wavelengths, which scatter less in tissue and thus result in much greater resolution and clarity of imaging than other technologies can deliver for the same vascular structures. The reduced scattering and much improved tissue penetration depth of NIR-II fluorescence photons made it possible for Hong to visualize deep brain structures in rodent animals.

After receiving his Ph.D. in chemistry from Stanford, Hong moved to the East coast, joining the Harvard lab of Professor Charles M. Lieber, one of the pioneers in nanoscience and nanotechnology. There, he worked on chronic brain and retina electrophysiology, using ultraflexible nanoelectronics that he developed. Since the early 2000s, Lieber had been using low-dimensional nanomaterials, such as silicon and other semiconductor nanowires, for biological applications, including biochemical sensing and electrophysiological measurement in cells and live animals. Hong had become fascinated by the research in the Lieber lab when he was still a graduate student at Stanford, and he hoped to change his research focus from optical nanomaterials to materials with unique electronic properties and potential for biological applications. He joined the project in the Lieber lab on the development of a new form of nanoelectronics, called 'mesh electronics', with brain-tissue-like ultraflexibility that makes them fundamentally different from conventional electronics fabricated with rigid materials. The mesh electronics devices are so soft and flexible that they can float on water like colloids and can be loaded into and injected out of a syringe like pharmaceuticals. Working with his colleagues in the Lieber lab, Hong demonstrated precise delivery of mesh electronics into any part of the brain and to other soft tissues, similar to therapeutic delivery of drugs.

Hong's return to Stanford as a faculty member was driven by his view that Stanford is the ideal place for him to continue his career. He cites the vibrancy of the scientific research as well as the welcoming atmosphere created by the people here. His lab is dedicated to a broad spectrum of interdisciplinary research at the interface between materials science and neuroscience, the science of the mind. Central to the research in his lab is the development of novel photonic and electronic materials that interrogate and modulate brain activity by different forms of energy (such as electricity, magnetism, light, sound and heat) in a minimally invasive manner. As his research progresses and unfolds in the lab, he imagines neuroengineering breakthroughs in the future that will enable reading and writing the neural code, as well as treating neurological diseases and even augmenting human intelligence via a non-invasive interface based on advances in materials science and engineering.

Hong has already received a number of awards in his career. He received a Graduate Student Award from the Materials Research Society in 2014 and was a recipient of the NIH Pathway to Independence Award, for the period 2017-2021.

Chapter 25

Appointments in Materials Physics

It has always been recognized that the underlying physics of matter provides a basis for understanding the structure and properties of materials. This was explicitly acknowledged in Dorsey Lyon's course in the Geology and Mining department in 1903-04 entitled "Metallography and Physics of the Metals." That understanding had not changed by the second decade of the 21st century when several faculty members were appointed who would focus on how and why the underlying physics of matter plays a key role in determining material behavior.

Aaron M. Lindenberg ~ Ultrafast Properties of Materials

Aaron Lindenberg joined the MSE department in 2007, with a joint appointment in the Department of Photon Science at the Stanford Linear Accelerator Center (SLAC). His appointment was the result of a joint search between the MSE Department and the Stanford Synchrotron Radiation Laboratory (SSRL): an appointment that had been urged by a department visiting committee in late 2002. The visiting committee felt strongly that MSE could make an immediate and potentially revolutionary impact in nanomaterials research by strengthening its ties with SSRL and by exploiting the remarkable x-ray capability that was being built at that facility. The committee specifically mentioned the Sub-Picosecond Pulsed Source (SPPS) and Linac Coherent Light Source (LCLS) as facilities that could permit MSE to become a world leader in time-resolved structural studies and characterization of both hard and soft materials and biomaterials on unprecedented time and length scales. Much of this promise was realized with the appointment of Aaron Lindenberg, who would build a strong research program in ultrafast properties of materials and earn tenure for that work.

Lindenberg had been a staff scientist at SLAC since 2003 after receiving his Ph.D. from UC Berkeley in 2001. Lindenberg's work has been focused on the ultrafast properties of materials, with the goal of visualizing, capturing, and directing the atomic-scale processes that underlie how materials and devices function. This work has broad applications to information storage technologies, energy-related materials, and nanoscale optoelectronic devices. He has developed and taught two new courses since joining the department, an introductory course on quantum mechanics for undergraduate engineering majors, and a graduate level course on advanced x-ray science and techniques. Since joining the MSE faculty, Lindenberg has built a strong

Aaron Lindenberg, 2012. (MSE Dept.)

research group that now includes five post-doctoral fellows and five graduate students.

Lindenberg was born in State College, Pennsylvania, on April 27, 1974. His father is in the computer business, but was initially a mathematician with a Ph.D. from the University of Oregon. His mother is a nutritionist. Lindenberg grew up in Sudbury, Massachusetts, just outside of Boston, where he graduated from Lincoln-Sudbury Regional High School in 1992. He then attended Columbia University, where he majored in physics, graduating with a B.A. degree in 1996.

Lindenberg moved to UC Berkeley for graduate work in physics. There he pursued his Ph.D. degree with his advisor, Roger Falcone. His main work was at the Advanced Light Source (ALS) at Lawrence Berkeley National Laboratory, where he carried out some of the first time-resolved x-ray studies of materials, visualizing how phase transitions occur on few picosecond timescales. After he received his Ph.D., he stayed on at Berkeley for about a year and a half as a faculty fellow, during which time he continued these time-resolved measurements and also taught several classes, including courses on statistical mechanics and on the physics of music and musical instruments.

For post-doctoral work, Lindenberg moved to SLAC in 2003, where he was a staff scientist until 2007. There he was involved in the early stage efforts to develop the first x-ray free electron laser, the Linac Coherent Light Source (LCLS), and carried out some of the first femtosecond resolution experiments at this source and precursors to it.

Lindenberg says that after joining the MSE Department in mid-2007, he found, through his interactions with his new colleagues, many new research opportunities and the world of materials science started to open up to him. In recent years, his group has been investigating new ways of using light to manipulate and engineer the properties of materials, including their structural, opto-electronic, and topological aspects. He has also developed a new understanding of how dynamical processes in materials impact their functionality. His most exciting recent work involves experiments which show that interlayer shear excitations in layered two-dimensional materials can be used to control their topological properties on ultrafast time scales (*Nature*, 2019).

Lindenberg developed two new courses in his first years at Stanford and has been teaching them ever since. The first is an introductory course on quantum mechanics for undergraduate engineering majors, "Quantum Mechanics of Nanoscale Materials," which is broadly focused on understanding materials properties from the perspective of quantum mechanics. The second is a more advanced graduate level course, "X-Ray Science and Techniques," which is designed for students engaged in research involving x-ray techniques at synchrotrons, free electron lasers, or laboratory-based sources. This course strongly overlaps with his research and provides a broad introduction to fundamental principles of x-ray scattering; spectroscopy, and photoemission; experimental techniques; and new accelerator-based sources.

Since joining the faculty, Lindenberg has received a number of awards for his research.

These include: a Terman Fellowship (2007-2009), a Department of Energy Outstanding Mentorship Award (2008), a DARPA Young Faculty Award (2010—2012), and a Chambers Fellowship at Stanford (2015—2018).

Jennifer A. Dionne ~ Nanophotonics

Jennifer Dionne won an MSE faculty search in 2008, even before she had finished her Ph.D. degree in applied physics at Caltech. She had worked with Harry Atwater at Caltech on nanophotonics, the interaction of light with matter at feature sizes smaller than the diffraction limit, not long after Mark Brongersma had been there as a post-doc working in the same general area. That Dionne was selected for the appointment even though Brongersma already had a successful department program on nanophotonics speaks to the very high regard the selection committee and the MSE faculty have for her. That she was permitted to spend a year as a post-doc with Paul Alivisatos at Berkeley following her graduation from Caltech in 2009, but before arriving at Stanford in 2010, is another measure of the confidence the department has in her.

Jennifer A. Dionne, 2012. (Jennifer Dionne)

The promise that she showed was quickly realized, as she soon developed an active and successful research group at Stanford. She has also become a popular teacher. She has received many honors for her research work at Stanford including the Materials Research Society Outstanding Young Investigator Award in 2017. Dionne was only the second woman to be appointed to the MSE faculty.

Jennifer Dionne was born in Warwick, Rhode Island, on October 28, 1981. Her father worked in construction with a specialty in cabinetry and her mother was a nurse. She was schooled at St. Mary Academy—Bay View, the all-girl Catholic School located in Riverside, Rhode Island. After graduating from high school in 1999, she attended Washington University in St. Louis, where she studied both physics and systems science and mathematics, receiving dual B.S. degrees in those subjects in 2003. Following her graduation from Washington University, she enrolled at Caltech in Pasadena for graduate work in Applied Physics, receiving an M.S. degree in 2005 and Ph.D. in 2009. It was at Caltech that she was introduced to the field of nanophotonics through her graduate work with Harry Atwater.

One of Dionne's earliest papers was published in *Science* when she was just 25. It was a paper on negative refraction of visible light, which she published with Henri Lezec and Harry Atwater, the latter then her advisor. The paper described how a negative-index material could be created by fabricating an ultrathin gold-silicon-nitride-silver (Au-Si_3N_4-Ag) structure, which acted as a waveguide for sustaining surface plasmons in the two metals, causing a negative index of refraction to be observed. This was the first time a negative index of refraction had been created for visible light. It launched her career in nanophotonics, and is now among her most highly cited papers.

Later at Caltech she published another of her highly cited papers, this one on a device she called a "plasmoster," a plasmonic device that converts optical signals into surface electromagnetic waves propagating along metal-dielectric interfaces. She showed that it functions like the metal-oxide semiconductor field effect transistor (MOSFET), which is the basis of modern computer chips. She pointed out that because surface plasmons exhibit extremely small wavelengths and high local field intensities, the optical confinement can scale to deep sub-wavelength dimensions in plasmonic structures, which then have the potential for opto-electronic and perhaps even all optical Si-based modulation. She followed up on this work during her post-doctoral studies with Paul Alivisatos at Berkeley, where she published papers on plasmonic architectures for integrating photonics into silicon chips. With Paul Alivisatos, she also began branching out into photochemistry, laying the groundwork for a field now known as plasmon photocatalysis.

At Stanford, Dionne has continued to focus on plasmonics, involving the measurement and manipulation of collective excitations of free electrons in nanoscale materials, and, more generally, on nanophotonics. She has studied both fundamental aspects of these topics and their applications in devices, such as solar cells, and in the characterization of nanoparticles and biological samples. High resolution electron microscopy and spectroscopic probes employed in the electron microscope have been major tools in her group's research. Indeed, in her second most highly cited work, published as the cover story in *Nature* (2012) with Jonathan Scholl and Ai Leen Koh, she made use of Scanning Transmission Electron Microscopy (STEM) to study the quantum plasmon resonances of sub-10nm metallic nanoparticles. With these same authors and postdoc Aitzol Garcia Etzarri, she followed this work with a *Nano Letters* 2013 publication revealing electron tunneling between two closely spaced metallic nanoparticles. This work promises to have important applications including single-molecule sensing and spectroscopy, novel nanoantennas, molecular rulers, and nonlinear optical devices.

Dionne has also developed new experimental methods, including methods to bring light to transmission electron microscopy and techniques to visualize dynamic solute-driven phase transitions with nanometer-scale resolution. As a model system, she has focused on the hydrogen-induced alpha-to-beta phase transition in individual palladium nanocrystals. In contrast to ensemble measurements of nanoparticles, her results indicate that sub-30 nm single-crystalline nanoparticles do not exhibit phase-coexistence, and that surface effects dictate the size dependence of the hydrogen absorption pressures. She has also shown that intercalation always begins from particle corners, regardless of shape, but that this mechanism and its associated kinetics can be modified with plasmon-resonant illumination. This work may have applications in improving the efficiency of photocatalysts, and in improving the storage and cycling abilities of batteries, where visualizing such phase transitions is very challenging. Moreover, Dionne led the development of a new tomographic technique, cathodoluminescence tomography, to visualize light-matter interactions in 3D with nanometer-scale spatial and spectral resolution. This technique may find applications that include the identification of recombination centers in light emitting diodes and photovoltaics, and new label-free avenues for bioimaging.

In solar energy, the Dionne group is researching new upconverting materials that convert low energy photons into higher energy photons. This approach allows solar cells to utilize portions of the solar spectrum that would normally go to waste and may significantly improve the cost/efficiency balance of photovoltaic and photocatalytic devices. Further, in the realm of biology, Dionne is developing new nanomaterials to visualize and manipulate nano-sized specimens, especially proteins and small molecules. Her nano-materials control chiral light-matter interactions, enabling sensing, spectroscopy, and dynamic manipulation of chiral macro-molecules, and offer the potential for all-optical separation of small chiral molecules in pharmaceuticals and agrochemicals. For this work, she was recently named a Moore Inventor Fellow.

Dionne has taken her turn in teaching in the MSE core, including "Materials Chemistry" (MatSci 202), and "Waves and Diffraction" (MatSci 205). In addition, she has taught the introductory course, "Principles of Electronic/Opto-electronic Materials and Devices" (MatSci 152) involving subjects close to her research experience. In an effort to promote the field of materials science and engineering at Stanford, Dionne created and has taught a freshman seminar course , "Science of the Impossible" (MatSci 82N), where she asks her students to "Imagine a world where cancer is cured with light, objects can be made invisible, and teleportation is allowed through space and time," which, she says, is "becoming a reality, thanks to advances in materials science and engineering."

Dionne's already long list of awards since joining the MSE faculty include: an Air Force Office of Scientific Research Young Investigator Program Award (2010); an Outstanding Young Alumni Award, Washington University in St. Louis (2012); a National Science Foundation CAREER Award (2012); a Presidential Early Career Award in Science and Engineering (2014); and the Materials Research Society Outstanding Young Investigator Award (2017); among many others.

When Dionne is not in the lab, you can find her exploring the intersection of art and science, cycling the latest century (100-mile) cycling course, and reliving her childhood with her husband Nhat and her two young sons, Marcus (now 4) and Hugo (now 2).

Evan J. Reed ~ Bringing Computational Materials Science to the Department

Evan Reed joined the faculty in 2010 as assistant professor to help grow the department's activities in computational approaches to materials science. He introduced the department's first graduate course focused on computational methods, "Atomistic Computational Methods for Materials" (MSE 331). This class is taken by graduate students in MSE and other departments across the School of Engineering, as well as the departments of physics and chemistry who are interested in employing these methods in their own research. He also introduced, "Nanoscale Materials Physics Computation Laboratory" (MSE 165/175), an undergraduate/graduate course enabling students to "discover" fundamental principles of materials using interactive computer simulations. He took over responsibility for, "Atomic Arrangements in Solids" (MSE 203) from Robert Sinclair in 2010.

Reed's research group at Stanford has elucidated some of the properties of so-called two-dimensional materials like graphene. His group predicted that some of these materials exhibit piezoelectric effects and can be

Evan Reed, 2010. (Evan Reed)

employed as structural phase change materials for electronic applications, both later observed in the laboratory. His most highly cited papers are based on that work. More recently, his group has explored the potential for machine learning and data mining to be useful for materials science challenges. Applications in addition to 2D materials include materials selection for lithium ion batteries and predicting the evolution of complex solid-state chemical reactions with thousands of reactions.

Reed was born in a rural area of Minnesota and exhibited early interests in aviation and space. He earned his pilot's license at 18, an event that would circuitously lead him to Stanford some years later. Reed attended Caltech, majoring in applied physics, and graduating in 1998. At Caltech, Reed worked with Professor B. Vincent McKoy and colleagues to develop a code that computes integrals central to calculation of electron scattering cross sections of molecules. This was a formative experience that built his interest in computational approaches to materials.

Reed earned his Ph.D. in physics at MIT in 2003 on shock compression of materials with Professor John Joannopoulos. His Ph.D. thesis includes predictions of reversed and anomalous Doppler effects that can occur when light reflects from a shock wave propagating through a photonic crystal. At MIT, Reed was exposed to a style of doing research that combines creativity with technical innovation, a style that he has aimed to emulate at Stanford.

After a short postdoc at MIT with John Joannopoulos, Reed moved to Lawrence Livermore National Laboratory (LLNL) from 2004 to 2009 where he was an E. O. Lawrence Postdoctoral Fellow and later a staff scientist working with Laurence E. Fried and colleagues. There he studied shock compression of energetic materials, making the first calculations of the complex chemistry and electronic structure that occurs behind the shock front during detonation. This work shed light on the microscopic chemical origins of the stability variation of various energetic materials to mechanical perturbations, like being accidentally dropped onto a floor. These studies were enabled by a multi-scale modeling method Reed developed with colleagues there. They also used the method to discover that the dirty ices that make up comets can chemically react to form amino-acid precursors when compressed by a shock wave, suggesting the building blocks of life may be more prevalent and robust than previously appreciated.

At Livermore, Reed saw how applied engineering research is done and what distinguishes approaches that have value and those that do not. This experience was complimentary to his MIT experience, and later enabled him to have more impact in applied fields at Stanford.

Reed has been a Robert Noyce Faculty Scholar within the School of Engineering at Stanford and a recipient of the DARPA Young Faculty Award, an NSF CAREER Award, and an Office of Naval Research Young Investigator (YIP) Award. He was promoted to the rank of tenured Associate Professor in 2018.

While at LLNL, Reed was active as a flight instructor at the Livermore airport, teaching others to fly small airplanes and to qualify for FAA certification. Through this process he discovered his own interest in teaching and mentoring, and began to consider the possibility of moving to an academic job, leading to his arrival at Stanford in 2010. Reed continues to fly out of the Palo Alto airport.

Persis Drell ~ From Atomic Physics to University Leadership

Persis S. Drell joined the Stanford faculty as a professor and director of research of the SLAC National Accelerator Laboratory in 2002. In 2004 she took on the role of deputy project manager for the Large Area Telescope (LAT), the main instrument of the Gamma-Ray Large Area Telescope (GLAST) then under construction at the lab. She was named SLAC deputy laboratory director in May of 2005, with responsibility for the particle physics and astrophysics research areas. In December 2007 Drell was named the laboratory's fourth director, a post that she held until 2012. During her time at SLAC, the laboratory gradually changed its emphasis from research on high-energy physics to a broader research agenda involving many different scientific disciplines, including materials science. In 2010, SLAC initiated the Linac Coherent Light Source (LCLS), the most powerful x-ray free electron laser in the world, and its research program has led to major advances in such diverse fields as energy and environmental science, drug development and materials science and engineering.

Drell has been credited with broadening the focus of SLAC and with developing stronger connections between SLAC and research activities on the Stanford campus, including materials research. In 2014, she was appointed the Frederick Emmons Terman Dean of Engineering She was also appointed the first James and Anna Marie Spilker Professor in the School of Engineering, professor in materials science and engineering, and professor of physics. Her appointment in the MSE department recognized the strong connections that had developed between the directorates of SLAC and the MSE department during her time at the laboratory.

Drell studied mathematics and physics at Wellesley College before receiving a Ph.D. in atomic physics in 1983 from the University of California at Berkeley. She then worked for a time as a post-doctoral scientist in high-energy experimental physics at the Lawrence Berkeley Laboratory, before accepting an appointment to the physics faculty at Cornell in 1988. She joined the faculty at SLAC 14 years later.

As dean of engineering, Drell focused on long range planning and how the school should be configured to best address the opportunities and challenges of the future. That planning process led to the establishment of goals in research, education, and culture that the school would pursue. Drell placed considerable emphasis on diversity and inclusion by focusing on increasing the role of women and underrepresented minorities in engineering. Drell also provided strong support for multidisciplinary research initiatives and for faculty hiring via multi-department faculty

Persis Drell (Stanford News Service)

searches addressing topics of broad interest in engineering. For example, she launched the Catalyst program, which funds faculty research teams addressing "big problems" requiring a multidisciplinary approach. She also approved broad multi-department faculty searches on topics such as robotics, materials and chemistry for the environment, and quantum science and engineering. In 2017, she was named Stanford University's 13th Provost.

Drell has collected many honors in her career, including election as a member of the National Academy of Sciences and as a Fellow of both the American Academy of Arts and Sciences and the American Physical Society. Early in her career she was a recipient of a Guggenheim Fellowship and a National Science Foundation Presidential Young Investigator Award.

Part V ~ Closing Observations

Few, beyond perhaps the prescient Professor Dorsey Lyon, would have predicted that Lyon's 1903 course on metallography and the physics of metals in Stanford's Geology and Mining Department would lead to today's Department of Materials Science and Engineering, one of the premier materials science departments in the world.

Due to its close connection to mineral resources and the mining industry during its first decades, the department's focus was initially on metals and metallurgy, much as it was for most other academic programs in the U.S. that would similarly transition into departments of materials science and engineering. By the 1910s, this early emphasis on the processing of metals was augmented, as more attention became focused on the structure and properties of metallic alloys, especially steels. This metallurgical focus of the department held well into the 1950s, and its shifting relationships with geology and other academic departments are reflected in the various names of the department and in its academic affiliations during that period.

It was not until after the name of the department was formally changed to Materials Science, in 1960, that semiconductors and their optoelectronic properties came to be one of its main thrusts. Even so, the department would not develop a full embrace of materials issues related to silicon technology until the 1980s. Work on solid-state electrochemistry involving non-metallic, ionic conducting crystals that started in the late 1960s laid the foundation for the explosion of work on batteries, fuel cells, and electrochemical sensors in the early part of the 21st century. Except for Dietrich's report on clays and ceramics in the 1920s, work on ceramics, especially structural ceramics, had not been a significant focus of the department. But research on electronic ceramics would be a natural development when problems and opportunities associated with integrated circuits came to be addressed in the 1980s and 1990s.

Polymeric materials represent another major branch of materials science that was long neglected by the department on the grounds that such work was naturally developing in the departments of chemistry and chemical engineering. But that assumption became unjustifiable when it was realized that the structure-processing-properties paradigm that had guided work on crystalline metals and other, inorganic, crystalline materials was applicable to polymeric materials (plastics) as well. It also became apparent that some of the most exciting work in materials science now involved polymeric materials. Embracing polymeric and related materials and their applications characterized the rebuilding of the department in the 21st century, leading to an increase in the size of the MSE Department and enhancing its reputation. By this century's second decade, ultra-soft materials of the kind that occur in biology and medicine had been added to the array of subjects being addressed by the department, further enhancing its reputation.

The word "microstructure" appears in Professor Lyon's 1903 course description, Stanford's very first course in materials (rather than in "assaying" or simply "metallurgy"). Then, and for many decades thereafter, the optical metallurgical microscope was the main tool for characterizing microstructure. From the low power bench microscopes of the kind that had been used in the late 19th century, to the more sophisticated optical metallographs that were used by mid-20th century, optical microscopy was the main tool for characterizing microstructure. With the invention of the transmission electron microscope (TEM) in the 1940s, and its commercialization in the 1950s, a new tool for materials characterization became available in materials science. The department acquired its first electron microscope in the 1960s, but was well behind the leading work in this field being done in England. The department began to catch up in the 1970s just as TEM was emerging as a much more diversified characterization instrument, which included micro-chemical analysis, convergent beam electron diffraction, and electron energy loss spectroscopy, as well as high resolution, lattice imaging. Scanning electron microscopy, which is now a standard tool for microstructure characterization, was not widely available until the early 1970s; the first commercial scanning electron microscope appeared in 1965. Thus, by the 1970s, a much wider array of tools was available for characterizing the microstructure of materials. This has had a profound impact on progress in materials science at Stanford and elsewhere.

In tracing the arc of materials science at Stanford for more than a century, it is remarkable to note how slowly gender diversity has developed in our department and our field. We have far to go to achieve gender parity, particularly when simply compared to other engineering and applied sciences departments at Stanford. For the first four decades of the department (1919-59), only one female student graduated from the department. As late as 1959, the School of Mineral Sciences made clear in the Course Bulletin that it aimed to train "high caliber men for responsible positions in industry, government, and education." Not until the 1960s would the department make a solid beginning, attracting a very small number of talented women who received their M.S. degrees that decade. It was not until 1973, however, that the first woman was granted a Ph.D. by the department. While a few more women received their Ph.D.s during the 1970s and 1980s, the output was still just a trickle compared the number of men receiving the same degree. As of 2019, fewer than 20 percent of the doctoral graduates are women. Sadly, the record for faculty appointments is worse. It was not until 2006 that the first woman was appointed to a tenure-track position on the MSE faculty. Presently, only three faculty members out 17 are women.

The department can claim significant geographical diversity, however. Of the department's first 14 students to enroll in its new graduate program in 1919, two were international students: Takeo Shikamura, from Japan, and Hsaio Wu, from China. Since that time, students and, more recently, faculty have been drawn from around the world as well as from across North America.

Having studied this past century of materials science and engineering at Stanford, I think it appropriate to consider what might now be expected for its future. As in the past, the direction the department might take will likely flow from the interests of faculty members who are appointed. Thus, new faculty appointments essentially determine the future of the department, especially because the new appointees are now the ones who will have a say in future appointments.

There was a time when faculty appointments were made in areas considered to be essential to a well-rounded department teaching program. When your thermodynamics man, and it was a man in those times, retired, you looked for a replacement in thermodynamics, thinking that that was one of the main subjects that must be covered. Indeed, it was commonly assumed that the faculty billet that had been held by the leaving faculty member was "owned" by the department and would be used to find a replacement in that general area. Thus, just as metallurgist Galen Clevenger was replaced by metallurgist Welton Crook in the early years, when Craig Barrett, who had initiated work in electron microscopy at Stanford, left Stanford for Intel, his replacement was Robert Sinclair, who then brought electron microscopy at Stanford to a new level.

The "ownership" of faculty billets by the department began to fade, however, in the 1980s, when the policy and practice of returning vacant billets to the School of Engineering was adopted. Through that policy, the school was able to move resources from one department to another so that it could respond to important opportunities throughout engineering. From that time forward, all departments in the school could make the case for new faculty appointments each year, regardless of whether any faculty members from that department had left during that year. In this manner, the authorization of new faculty searches was based on the needs and opportunities that a department could best articulate. John Bravman was appointed in the 1980s because a good case was made that the department needed to pursue materials issues related to silicon technology. Bruce Clemens was appointed in the late 1980s in part to fill an important gap in kinetics created when Marsh Pound took medical retirement earlier in that decade. Reiner Dauskardt, Shan Wang, and Paul McIntyre were appointed in the 1990s to respond to perceived needs in structural materials, magnetic materials, and electronic ceramics, respectively. These appointments did not involve filling "department owned" billets per se but were based on the perceived and articulated needs of the department in certain core subjects in materials science.

Since the turn of the 21st century, faculty appointments in MSE have been made less to fill core needs of the department following the classical structure-properties-processing paradigm, and more to pursue promising research opportunities in other, newer fields where materials science can play a role. Many faculty in MSE appointed since that time have made their reputations by focusing on what one can do with materials in engineering or medical applications, rather than on how one can manipulate materials' properties or behavior by changing their structure though processing, as was done for most of the past century. Examples of this current work include the development of solar cells; batteries; smart windows; chemical and biological sensors; electronic, photonic and plasmonic devices; and dermatological and stem cell therapies. Thus, the focus of the department is now less on how materials are constituted and how they behave, than on how material behavior can be used to address certain pressing problems in technology, the environment, biology, and medicine.

As more and more faculty are appointed for their work on these subjects, the center of gravity of the department can be expected to shift further in that direction, leaving for other departments and universities other promising fundamental subjects in materials science and engineering such as high entropy alloys, intermetallic alloys and novel additive manufacturing.

Judging from this trajectory, it seems likely that the future MSE Department at Stanford will be characterized more by its work on major societal and technological problems and less for its work on the classical structure-properties-processing paradigm that defined the field for most of the last century.

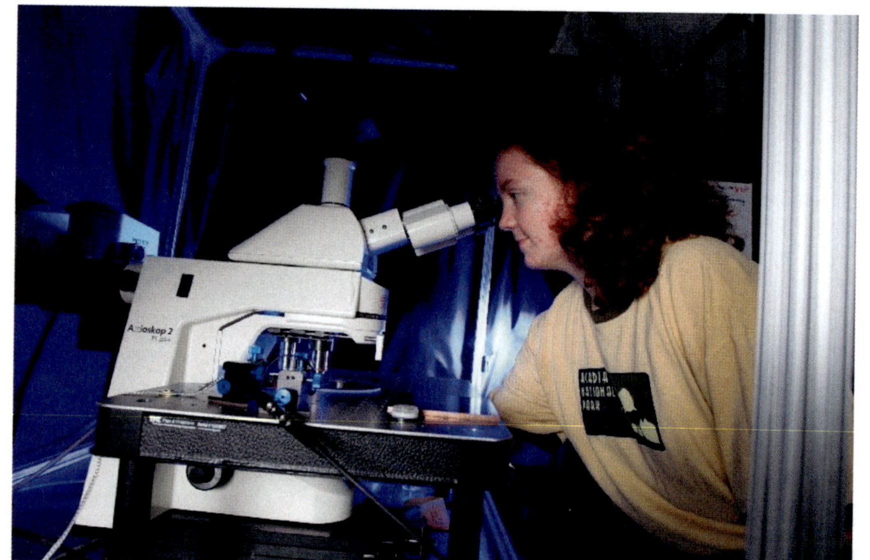

Student Elizabeth Hager-Barnard, 2006. (School of Engineering

Michael McGehee with student Tomas Lietjens, 2017. (Stanford News Service)

Student Markus Ong in Materials characterization lab, 2006. (School of Engineering)

List of Acronyms and Abbreviations

AAAS	American Association for the Advancement of Science
AAAS	American Academy of Arts and Sciences
A.B.	Artium Baccalaurei (undergraduate degree: later became B.A., Bachelor of Arts degree)
ACS	American Chemical Society
AEC	Atomic Energy Commission
AFOSR	Air Force Office of Scientific Research
AI	Artificial Intelligence
AIChE	American Institute of Chemical Engineers
AIME	American Institute of Mining, Metallurgical and Petroleum Engineers
ALS	Advanced Light Source (Berkeley)
APS	American Physical Society
ARPA	Advanced Research Projects Agency
ASM	American Society for Metals, later ASM International
ASTM	American Society for Testing and Materials
AT&T	American Telephone and Telegrapy
B.A.	Bachelor of Arts degree
BASc	Bachelor of Applied Science degree
BOSP	Bing Overseas Studies Program
B.S.	Bachelor of Science degree
Cal	University of California, Berkeley
Cal Poly	California Polytechnic School
Caltech	California Institute of Technology
CCMRD	Interagency Coordinating Committee on Materials Research and Development
CEO	Chief Executive Officer
CIS	Center for Integrated Systems, became SNF Stanford Nanofabrication Facility
CMR	Center for Materials Research, became Geballe Laboratory for Advanced Materials
CMN	Stanford Center for Magnetic Nanotechnology
CMOS	complementary metal-oxide-semiconductor
CMR	Center for Materials Research
CMU	Carnegie Mellon University
COO	Chief Operating Officer
CRISM	Center for Research on Information Storage Materials, became Stanford Center for Magnetic Nanotechnology
CT	computed tomography

CUSPEA	China-US Physics Examination and Application
C.V.	curriculum vitae
CWP	cold water pipe
d-school	Hasso Plattner Institute of Design at Stanford
DARPA	Defense Advanced Research Projects Agency
DOD	Department of Defense
DOE	Department of Energy
DRAM	dynamic random access memory
DSSC	Data Storage Systems Center
ECS	The Electrochemical Society
EE	Electrical Engineering
Eimac	Eitel-McCullough Company
EPASA	Electron Probe Analysis Society of America
EPRI	Electric Power Research Institute
ERDA	Energy Research and Development Administration
ESCA	electron spectroscopy for chemical analysis
FaAA	Failure Analysis Associates
FDA	Food and Drug Administration
FRS	Fellow of the Royal Society
GE	General Electric
G.I. Bill	Servicemen's Readjustment Act of 1944
GLAST	Gamma-Ray Large Area Telescope
GMR	General Motors Research
GMR	giant magnetoresistance
GPA	grade-point average
HRL	Hughes Research Lab
HRTEM	high-resolution transmission electron microscopy
HVSEM	High-voltage scanning electron microscope
IBM	International Business Machines Corporation
ICP-MS	Inductively Coupled Plasma Mass Spectrometry
IDL	InterDisciplinary Laboratory / Laboratories
IEEE	Institute of Electrical and Electronics Engineers
INCO	International Nickel Company
ISI	Institute for Scientific Information
JUMP	Joint University Microelectronics Program
KAIST	Korea Advanced Institute of Science and Technology
KAUST	King Abdullah University of Science and Technology
KIST	Korea Institute of Science and Technology
KMSE	Korean MSE graduate student organization
LANL	Los Alamos National Laboratory
LAT	Large-Area Telescope

LCLS	Linac Coherent Light Source
LED	light-emitting diode
LLNL	Lawrence Livermore National Laboratory
MAC	Materials Analysis Company
MAF	Mineral Analysis Facility
MAS	Microbeam Analysis Society
MATMOD	materials model
MBE	molecular beam epitaxy
MIT	Massachusetts Institute of Technology
MOCVD	metal organic chemical vapor deposition
MOSFET	metal-oxide semiconductor field effect transistor
MNP	magnetic nanoparticles
MOSFET	metal-oxide semiconductor field effect transistor
MRAM	magnetoresistive random access memory
MRL	Materials Research Laboratory program
MRS	Materials Research Society
M.S.	Master of Science degree
MSA	Electron Microscopy Society of America
MSE	Materials Science and Engineering
MTJ	magnetic tunnel junction
NAE	National Academy of Engineering
NAF	Mineral Analysis Facility
NAS	National Academy of Sciences
NASA	National Aeronautics and Space Administration
NATO	North Atlantic Treaty Organization
NBSE	National Black Society for Engineers
NIR-II	second near infra-red
NNCI	National Nanotechnology Coordinating Infrastructure
NSA	National Security Agency
NSF	National Science Foundation
ONR	Office of Naval Research
OSTP	White House Office of Science and Technology Policy
OTEC	Ocean Thermal Energy Conversion
PARC	Palo Alto Research Center (Xerox)
PEP	Positron Electron Project
Ph.D.	Doctor of Philosophy degree
PSAC	President's Science Advisory Committee
PVMI	Photovoltaic Manufacturing Initiative
PWA	Pratt & Whitney Aircraft
PYI	Presidential Young Investigator award
RCA	Radio Corporation of America

SBSE	Society of Black Scientists and Engineers
Sc.B.	Bachelor of Science degree
Sc.D.	Doctor of Science degree
SCMN	Stanford Center for Magnetic Nanotechnology
SEM	Scanning Electron Microscope
SLAC	Stanford Linear Accelerator Center (today's SLAC National Accelerator Laboratory)
SNF	Stanford Nanofabrication Facility
SNL	Stanford Nanocharacterization Laboratory
SNSF	Stanford Nano Shared Facilities
SOT-MRAM	Spin-orbit torque magnetoresistive random access memory
SPEAR	Stanford Positron Electron Asymmetric Ring
SPIE	professional society for optics and photonics technology (formerly the Society of Photographic Instrumentation Engineers, later the Society of Photo-Optical Instrumentation Engineers)
SRC	Semiconductor Research Corporation
SRI	Stanford Research Institute
SSRL	Stanford Synchrotron Radiation Laboratory / Lightsource
SSRP	Stanford Synchrotron Radiation Project
STEM	Scanning Transmission Electron Microscopy
STEM	Science, Technology, Engineering and Mathematics
SWCNTs	single-walled carbon nanotubes
TEM	transmission electron microscopy
TI	Texas Instruments
TMR	tunneling magnetoresistance
TMS	The Minerals, Metals & Materials Society
UBC	University of British Columbia
UC	University of California
UCLA	University of California Los Angeles
UCSB	University of California Santa Barbara
USC	University of Southern California
USTC	University of Science and Technology (China)
UW	University of Wisconsin
VARTM	Vacuum Assisted Resin Transfer Molding
VG	vacuum generators
WEST	Women in Engineering, Science, and Technology
WGLN	Wallenberg Global Learning Network
WIN	Western Institute of Nanoelectronics
WRL	Wallenberg Research Link
XPS	X-ray photoelectron spectroscopy

Endnotes

Page 10 Hoover, 1925.
Theodore J. Hoover to Ray Lyman Wilbur, letter, April 17, 1925, in School of Engineering Records, SC0165, series I, box 6, folder 21, also quoted in C. Stewart Gillmor, *Fred Terman at Stanford* (Stanford: Stanford University Press, 2004) 74.

Page 11-12 Hoover, 1919.
Theodore J. Hoover to Ray Lyman Wilbur, letter, February 27, 1919, in School of Engineering, Deans Office Records, SC0165, series 1, box 5, folder 6, Stanford University Archives.

Page 30 Parkman, 2007.
Ralph Parkman to William D. Nix, letter, 2007, in author's collection.

Page 47, 48 NAS, 2017.
Oleg Dimitri Sherby, by J. Wadsworth and W. D. Nix in *National Academy of Engineering of the United States of Americax: Memorial Tributes*, 21 (2017) 374-392.

Page 50 Nobel, 2014.
Press Release of the Nobel Prize in Physics, 2014, available at http://www.nobelprize.org/prizes/physics/2014/summary (accessed September 16, 2019)

Page 51 Maruska, 2015.
H. P. Maruska and W. C. Rhines, "A Modern Perspective on the History of Semiconductor Nitride Blue Light Sources," *Solid-State Electronics*, 111 (2015) 32-41.

Page 54 Pettit, 1963.
Text of Joseph Pettit Speech at Dedication of Peterson Building, in School of Engineering Dean's Office Records, SC0165, series 1, box 1, folder 19, Stanford University Archives.

Page 61-62 Tiller, n.d.
Tiller, William A., "An Extended Bio of William A Tiller and His Psychoenergetics Research," (undated), sent to the author, February 16, 2017.

Page 74 Tetelman, 1980.
Memorial Resolution for Alan Stephen Tetelman, Materials Science Department, UCLA, prepared for the Academic Senate of the University of California, by Kani Ono, Christian N. J. Wagner, Russell A. Westmann, and Alan Ardell, 1980.

Page 113-14   Hughes, 2017.
Darcy Hughes to W. D. Nix, email, August 28, 2017, in author's collection.

Page 146   Spicer, 2004.
"William E, Spicer, Emeritus Professor of Engineering, Dies at 74," *Stanford Report*, July 1, 2004.

Chuck Packer, a student of Oleg Sherby, admiring a superplastically deformed tensile sample of AlZn from his research, 1960s. (Stanford News Service)

Bibliography

In addition to biographical statements provided to the author by MSE faculty members (as noted in the preface) and preserved in the William D. Nix collection, the author found the following sources useful:

Published Materials (articles and books)

"Alan Stephen Tetelman, Materials Science Department, UCLA," University of California, Academic Senate, 1980.

Bube, Richard H., *One Whole Life* (Privately published, 1995).

Elliott, Orrin Leslie, Stanford University, the First Twenty-Five Years (Stanford: Stanford University Press, 1937).

"Galen H. Clevenger," *Engineering and Mining Journal,* December 18, 1920.

Geballe, Theodore H., "Why I haven't Retired," *Annual Review of Condensed Matter Physics,* 4 (2013) 1-21.

Gillmor, C. Stewart. *Fred Terman at Stanford: Building a Discipline, a University, and Silicon Valley* (Stanford: Stanford University Press, 2004).

Hoover, Theodore J., "Training the Engineers of Tomorrow," *Stanford Illustrated Review* 27:8 (May 1926) 430-33.

L.eslie, Stewart W., *The Cold War and American Science: The Military-Industrial-Academic Complex at MIT and Stanford* (New York: Columbia University Press, 1994) Chapter 8: "Materials Science."

Maruska, H.P., and W.C. Rhines, "A Modern Perspective on the History of Semiconductor Nitride Blue Light Sources," *Solid-State Electronics,* 111 (2015) 32-41.

"Metallurgists of Note: Galen H. Clevenger," *Engineering and Mining Journal,* December 18, 1930, 1117.

"Metallurgy at Stanford," (Stanford University, 1957).

Minerals Separation Ltd. And Minerals Separation American Syndicate v. James M. Hyde, US Circuit Court of Appeals, 242 U.S. 261, 37 S.Ct. 82.

Mitchell, J. Pearce, Stanford University, 1916-1941 (Stanford: Stanford University Press, 1958).

Nash, George H. *Herbert Hoover and Stanford University* (Stanford: Hoover Institution Press, 1988.

"Oleg D. Sherby, Professor of materials science and engineering, dies at 90," Stanford News, January 8, 2016.

National Research Council. *Advancing Materials Research.* (Washington, DC: The National Academies Press, 1987).

Rickard, Thomas *A History of American Mining*. New York: McGraw-Hill. pp. 406–407 (1932).

Smith, James Perrin, "The Stanford School of Mines," *Stanford Alumni Review* 20 (April 1919) 332-22.

"Theodore Hoover is Made Head of New Department," *Stanford Daily*, April 19, 1919

"W. F. Dietrich," *Mineral Information Services* 5 (1959) 11.

Wadsworth, J., and W. D. Nix, "Oleg Dimitri Sherby," *National Academy of Engineering of the United States of America: Memorial Tributes* (Washington, D.C., National Academies Press, 2017).

William Spicer, Emeritus Professor of Engineering, Dies at 74," Stanford Report, July 1, 2004.

Wilbur, Ray Lyman, Memoirs of Ray Lyman Wilbur, 1875-1949. Edgar Eugene Robinson and Paul C. Edwards, eds. (Stanford: Stanford University Press, 1960).

Wooten, Paul, "Metallurgists of Note: Dorsey A. Lyon," *Engineering and Mining Journal,* November 20, 1920, 997.

Stanford University Publications

Campus Report (later, *Stanford Report).*

Stanford Alumni Review, Stanford Illustrated Review, and *Stanford Review* [alumni monthly magazine].

Stanford Daily [student newspaper, 1892-present].

Stanford University. Annual Register (1891-1947), later, *Courses and Degrees* (1948-).

Stanford University. Annual Report of the President (1903-1948).

Stanford University. Directory of Students and Faculty.

Stanford University, Academic Council. [Faculty] Memorial Resolutions (Academic Secretary's Office).

Unpublished works of note

Crook, Welton J. "A Partial History of Metallurgy at Stanford University," unpublished symposium paper, 1968 (in author's collection).

Elliott, John E., to Herbert Hoover Medal Committee, Stanford University, 1972, Materials Science and Engineering Department faculty file (Welton Crook), copy in author's collection.

Geballe, Theodore H., "Family, Friends and Physics," (privately published, 2016).

Hoover, Theodore J., to Ray Lyman Wilbur, letter, 27 February 1919, in School of Engineering, Deans Office Records, SC0165, series 1, box 5, folder 6, Stanford University Archives.

Hoover, Theodore J., to Ray Lyman Wilbur, letter, April 17, 1925, in School of Engineering Records, SC0165, series 1, box 5, folder 21, Stanford University Archives.

Huggins, Robert, "The Evolution of Materials Science at Stanford," unpublished symposium paper, 1968 (in author's collection).

"List of graduate theses granted by the department from 1919 to 1949," Materials Science and Engineering Department, Stanford University (in author's collection).

Nix, William D., "The Evolution of Metallurgy at Stanford," unpublished symposium paper, 1968 (in author's collection).

Nix, William D., [the history of metallurgy and materials science], unpublished symposium paper, 1992 (in author's collection).

Shepard, O. Cutler, "Historical Sketch of the Department of Metallurgy and Materials Science," Dept. of Materials Science, Stanford University, 1992, with autobiography (brochure).

Terman, Frederick E., to Ivor Irving Levorsen, June 15, 1946, in School of Engineering, Deans Office Records, SC0165, series 1, box 5, folder 9, Stanford University Archives.

Tiller, William A., "The Future of Materials Science at Stanford," unpublished symposium paper, 1968.

Tiller, William A., "An Extended Bio of William A Tiller and His Psychoenergetics Research," (undated), sent to the author, February 16, 2017.

Tiller, William A., [The growth and development of the department in the 1960s and 70s], unpublished symposium paper, 1992.

"Turkish Students 1938-1944," School of Engineering, Dean's Office Records, SC0165, box 9, folder 27, Stanford University Archives.

Archival and Manuscript Collections, Stanford University Archives

Branner, John Casper, Presidential Papers, SC0065

Hoover, Theodore J., "The Stanford Years: Being an Excerpt from 'Memoranda, the Unpublished Autobiography of Theodore J. Hoover," SCM0231

Peninsula Times-Tribune Photographs of Stanford University, SC0065

School of Earth Sciences, Records, SC0405

School of Engineering, Dean's Office Records, SC0165

Stanford Historical Photograph Collection, SC1071

Stanford News and Publications Service Records, 1954-2010, SC0122 [biographical files]

Stanford Plant Services. Department Records, SC0123

Stanford University. Biographical Files (reference collection), SC1136

Sterling, J. E. Wallace, Presidential Papers, SC0216

Terman, Frederick E., Papers, SC0160

Tresidder, Donald B., Presidential Papers, SC0151

Wilbur, Ray Lyman, Presidential Papers, 1916-1943, SC064a

Interviews

Barrett, Craig R. Oral History for the International Archives of the Computerworld Honors Program, 2002. (Prepared by Daniel D. Morrow).

Bates, Clayton W., Jr. "Interview with Clayton W. Bates, Jr., March 5, 2004," The HistoryMakers African American Video Oral History Collection, Chicago, Illinois.

Bienenstock, Arthur L. "Arthur (Artie) L. Bienenstock, interviewed by Allison Tracy, February 19 and March 25, 2014," Stanford Oral History Project, SC0932, Stanford University Archives.

White, Robert L. "Robert L. White," interview, 2015, Stanford Oral History Project, SC0932, Stanford University Archives.

Stanford University, 2011. (Stanford News Service)

Index

Abernathy, Cammy R. 108
Abraham, Farid, 157
Ahlquist, Norman, *66*, *120*
Almassy, Marcia (Yamada), 110
Ahn, Byung-Tae, *127*,
Ahn, Channing, 173
Ahn, Kyung-Soo, *121*, 125
Amedei, Leo, *120*
Amine, Khalil, 157
Anderson, Jim, *56*
Akasaki, Isamu, 50
Alivisatos, Paul, 206, 221, 222
Amelinckx, Severin, 99
Amano, Hiroshi, 50
Appel, Eric, 211, 214-216
Ardell, Alan, 24, 73, 83, 107
Argon, Ali, 154,
Arnould, Helene F. (Netillard), 110
Asaro, Robert J. 59, *120*
Atwater, Harry, 194, 196, 221
Auld, Bertram, 139

Bacon, David, 95
Baer, Eric, 105
Bahney, Luther, 6, 19, 21
Baker, Shefford P. 106
Banke, Kirsten, 110
Bao, Zhenan, 201, 207,
Barbee, Troy, 152
Barnett, David M., 63, 64, 69, 81, 91-96, 99, 116, 121, 165, 166, 189
Barrett, Charles S. 98
Barrett, Craig Radford, 47, 57, 69, 74, 80, 81-85, 98, 106, 108, 110, 115-117, 119, 124, 129, 133, 229,
Bates, Clayton Wilson, Jr. 115-117, 129-130-132, 152
Bates, Priscilla, 131, 132
Beal, Carl H. 13
Beasley, Mac, 165, 172
Beaudry, Chuck, *32*
Benzunce, Jose, *118*,
Berla, Lucas A. 80
Beyers, Robert B. 106
Bienenstock, Arthur Irwin, 57, 60, 76, 81, 86-91, 129, 136, 165, 173
Bing, Peter S. 166
Bilir, Necmi, 124
Birringer, Ryan R. 80
Boehm, Alexandria, 307
Boukamp, Bernard, 41
Bourell, David L. *76*, 107
Bravman, John C. 106, 107, 159, 160, 163-167, 175, 178, 181, 229,
Branner, John Casper, 3-4, 13, 20, 22
Brennan, Sean, *122*, 120, 173
Briggs, Janet Zaph, 13, 109, 114
Brimhall, John L., 33
Brock, Ryan, 174
Brongersma, Mark, 194-196, 198, 203, 207, 221,
Brooks, Harvey, 89
Brotzen, Franz R. 92
Brown, James, 171
Brown, Walter, 195
Bube, Richard Howard, 57, 62-65, 70, 124, 134, 145, 152, 166, 167, 175, *176*, *177*, 191
Buccholz, Jeff, 172
Bustamante, Carlos, 156
Burmeister, Robert A. Jr., 50
Byer, Robert, 139, 140

Cahn, Robert, 98
Cai, Wei, 70, 93
Caligiuri, Robert D. 80, 120
Campanella, Roy, 96
Capehart, Wes, 172
Carry, Glen, *118*
Carry, Paul, *118*

Casper, Gerhard, 167
Castaing, R. 52-53
Chalmers, Bruce, 58-59, 98
Chan, Candice, 43
Chidsey, Christopher, 190
Chien, Chiang, 160
Chin, Bryan, 108
Chipman, John, 28, 36
Chmelka, Brad, 197
Choi, Jae-Won, 125
Chou, Tsu-Wei, 107
Chu, Steven, 207, 208
Chueh, William, 108, 208-210
Cima, Michael, 189
Çiraci, Salim, 124
Clemens, Bruce M. 80, 106, 108, 135, 136, 170-175, 181, 194, 229,
Clevenger, Galen Howell, 6-8, 13, 19-21, 30, 229
Clinton, William Jefferson (President) 90
Coakley, Kevin, 193
Colella, Whitney, 155
Coquin, Jerry, *56*
Cramden, Ralph, 121
Criddle, Craig, 207
Crites, Becky, *122*
Crook, Welton J. 6, 13-14, 16-17, 19-24, 27, 33-34, 50 (Award), 123, 229
Cui, Yi, 43, 80, 108, 205-208

Daehn, Glenn, *68*, 106
Dai, Hongjie, 217
Dauskardt, Reinhold H. (Reiner), 80, 104, 124, 136, 173, 181, 184-187, 229
Depelchin, Jacques, 119, *120*,
Dietrich, Waldemar Fenn, 7. 9. 11, 13-14, 27-28, 227
DiMaggio, Joe, 150
Dionne, Jennifer A. 108, 111, 221-223
Dizio, Kathy, 196
Doan, Jonathan, 160, 174
Doerner, Mary, 69
Doniach, Sebastian, 87
Dorn, John, 37, 45-47, 141
Drell, Persis, 224-226
Driscoll, Judith, 153
Dubois, Gerard, 157
Dundurs, John, 92
Duran, Servet A., 13, 36, 123
Durand, William F., 2, 15, 18
Durand Building 2, 18

Earhart, Chris, 174
Earthman, Jim, *68*, 118, 120, *121*, *122*
Edwards, Glen, *120*, 171
Eesley, Gary, 172, 173
Eiselstein, Lawrence E. 80
Eisenhower, Dwight David (President), 38-39
Eisenhower administration, 89
Eitel, William, 55, 101
Elliot, Grant, *120*
Elliott, John E. (Brick), 13, 22
Eom, Chang-Beom, 107, 153
Erten, Mehmet Hayn, 29, 123
Escalante, Jaime, *92*
Eshelby, J. D. 91, 93
Eshoo, Anna (Rep.) 162
Etzarri, Aitzol Garcia, 222
Evans, Mike, *119*
Ewald, Paul, 89

Fahrenbruch, A. L., 64
Falcone, Roger, 220
Fan, Shanhui, 207
Fedai, 110
Fehrer, Fred, *119*
Feigelson, Robert S. 137-141
Fert, Albert, 149, 182
Fine, Morrie, 99
Fischer, Ian, 139
Fischer-Colbrie, Alice, *110*
Fish, John L., 15
Flinn, Paul A. 157, 159-160, 165
Florando, Jeff, *104*
Fontana, Robert E. 157
Fouts, Caroline J. 110
Frank, Charles, 32, 204
Fredrickson, Glenn, 197
French, Edward P., 31
Frey, Donald N. 71
Fried, Laurence E. 224
Friedel, Jacques, 74, 204
Fuchs, Henry, 142
Fuchs, Ken, *121*
Fultz, Brent, 173,
Fuoss, Paul, *76*, 120

Gabaly, Farid El, 210
Galofalini, Steve, *121*

Galt, John, 150
Gambhir, Sanjiv (Sam) 136, 183
Gardner, Donald, 159
Gasca-Neri, Rogelio, 119, *120*
Gaster, Richard, 183
Gatos, Harry C. 106
Gay, Jack, 172, 173
Geballe, Theodore H. 57, 124, 140, 145, 150-153, 161, 165, 172,
Geballe Laboratory 90, 161, 187
Geçkinli, Ayşe Emel (Fedai), 84, 110, 111, 123, 124
Gent, William, 210
Giauque, William, 151
Gibbons, Jim, 164, 167
Gibeling, Jeffrey C. 107, 174
Gilgenbach, Ron, 173
Gilman, Paul, *76*, 117, 120, 121
Glaeser, Andy, 204
Gleason, Jackie, 121
Goetzel, Claus G. 100
Goll, Jeff, *120*
Gomez, Mario, *124*
Goo, Edward, 106
Goodier, J. N. 92
Goods, Steve, *76*, 80
Greer, Julia R., 70
Gregory, Richard, *56*
Greninger, Alden B., 13, 26-28
Grossman, Arthur, 154
Grünberg, Peter, 149, 182
Gulden, May Ellen, 110
Gunier, A. 52
Gür, Turgut, 41, 108, 115-116 (photo credit), 116, 124, 157
Guy, Albert G. 99
Guyer, Eric P. 80
Haasen, Peter, 100, 172
Haba, Belgacem, 121, *122*
Hadley, C. S. 169
Hager-Barnard, Elizabeth, *230*
Hagström, Stig B. 79, 136, 143, 152, 161, 163, 166, 167-179, 175, 176,
Halicioglu, Timur, 157
Hall, Lewis Dana, 33
Han, Seung-Min, 121, *125*, 126, *127*, 128, 174
Hanagud, Sathya V. 78
Hanson, Ron, 120
Harris, Steven, 139
Harrison, Walter, 124, 172
Hasso Plattner Institute of Design 18
Hayes, Garrett J. 80
Hearst, George, 1
Heath, Jim, 197
Hedrick, James, 215
Heeger, Alan J. 193, 194
Heilshorn, Sarah, 110, 112, 156, 196, 211-214
Hellawell, Angus, 98
Hellstrom, Eric, *120*
Hemker, Kevin, *68*, 107, 118, 121
Hennessy, John, 167
Henry, Joseph, 49
Heremans, Joseph, 172
Herbst, Jan, 172
Hessler, Tom 82
Hirth, John P. 72, 76, 77, 99
Ho, Chun, 41, 42
Hoff, Nicholas J., 37
Hoffman, Drew, 121
Holbrook, John, *120*
Honeycombe, Robert, 98
Hong, Guosong, 211, 216-218
Hoover, Herbert, 3, 10
Hoover, Herbert Jr., 12
Hoover, Lou Henry, 22
Hoover, Theodore J., 1, 3, 9-16, 21-22,26-27
Howe, Henry Marion, 5-6
Howe, Julia Ward, 5-6
Howe, Samuel Gridley, 6
Hufnagel, Todd, 174
Huggins, Robert A., 18, 33-45, 47, 49, 55, 68, 70, 72-74, 84, 97, 100, 105-106, 108, 115-117, 123-124, 127-128, 137, 139, 145, 161, 177, 193, 207
Hughes, Darcy A., 33, 113-114, 148
Hussen, Guleid, *104*, 121,
Hyde, James M., 13-15
Im, Ho-Bin, 124, 127
Jack, K. H. 133
Jackson, Kenneth, 59, 106
Jaffee, Robert I. 107, 108, 157
Jaynes, Edwin T. 131
Jeffries, Zay, 20, 23, 28
Joannopoulos, John, 224
Johnson, Leon, 120

Johnson, W. Robert, 33, 119,
Johnson, William, 171-173
Jordan, David Starr, 3-4, 20
Jordan, Knight Starr, 20
Jordan, Payton, 83
Justi, Stefan, 164

Kader, Billie, 203
Kang, Ho-Kyu, 125
Karnthaler, Professor, 154
Kassner, Mike, *68*, 120, *121*, *122*
Kay, Eric, 106
Kayali, Eyüp Sabri, 124
Kays, William M. 72, 143
Keating, Kenneth L., 31, 33
Keene, Scott, *202*
Keith, Doug, 152
Kelly, Michael A. 157, 161-162, 168, 169
Kempen, Paul, 174
Keum, Dong-Wha, 125, *127*
Killian, James, 39
Kim, Dae-Yong, *125*
Kim, Doek-kee, *127*
Kim, Jungyup, *125*
Kim, Ki-Soo, *127*
Kim, Yong-Suk, *127*
Kingery, W. D. 138
Kino, Gordon, 139
Kirchner, Professor, 154
Kitabjian, Paul, *68*
Klein, M. O. 169
Kline, Joe, 193
Knuth, Donald, 165
Koh, Ai Leen, 222
Kotler, Gerald, 58, 59
Kröner, Ekkehart, 93
Kryder, Mark, 182
Kwon, Unoh, *125*

Lairson, Bruce, 153, 174
L'Ampheres, Maud, 20
Langer, Bob, 215,
Latta, Gordon, 92
Laudise, Robert A. 106
Lawrence, Jerry, *116*, 117
Leake, John, 132
Lee, Chang-Beum, 125
Lee, Dong-Ick, *125*
Lee, Jae Young, 124
Lee, Seok-Hee, 125
Lee, T. D. 181
Lee, Yong-Won, *125*
Lezec, Henri, 221
Lieber, Charles M., 206, 217, 218,
Leiberman, David, 99
Leibert, Bruce E., 33, *52*
Lenox, Lionel Redmond, 4
Levi, Kemal, 124
Levine, Harold, 92
Lewis, Jack, 31-32
Li, James C. M. 74
Li,Yiyang, 210
Liaw, Bor Yann, 108
Lietjens, Tomas, *230*
Lin, Jer-Hong, *76*
Lindau, Ingolf, 168, 169
Lindenberg, Aaron M. 219-220
Lintner, Professor 154
London, Blair, *118*,
Lorenzini, Robert, 110
Lothe, Jens, *77*, 94, 95, 99
Lund, Richard W. 80, *116*, 117
Lyman, Donald J. 101, 211,
Lyne, Henry, 19
Lyon, Dorsey Alfred, 3-6, 8, 28, 219, 227, 228
Lytton, Jack 83

MacDiarmid, Alan G. 194
Macres, Victor Gregory, 37-38, 45, 52-53, *56*, 57, 81, 84
Maddix, Bob, *119*
Mansfield, Mike (Senator), 43
Maradudin, Alexei A., 31-32, 94
Maradudin, Alexei P., 13, 31
Marieb, Thomas, 157
Marlor, Guy, *63*, 64
Marshall, Ann, 134, 135
Marton, Ladislas L (Bill), 28, 53
Maruska, Herbert Paul, 49-52
Mash, Donald R., 31, 100
Massalski, Ted, 98
Matlock, David, 69, 106, 107, 108
Matthias, Bernd, 151
Maxwell, Paul, *120*
Mayo, Merrilea J. 106, 113
McCarty, Kevin, 210

McCullough, Jack A. 55, (Building: 40-41, 45, 51, 83, 101, 102, 148
McDaniel, Tony, 210
McEvily, Arthur J. 98-99
McGehee, Michael, 109, 192-194, 201, 203, 207, 209, *230*
McIntyre, Paul, 80, 124, 128, 175, 181, 187-190, 229
McKoy, B. Vincent, 224
Mecartney, Martha, 113
Mehl, Robert Franklin, 66, 76-77
Meindl, Jim, 164
Melosh, Nicholas, 161, 196, 197-200, 203
Melville, Herman, 49
Menezes, Ricardo, 119
Meredith, Charlotte, 110
Meyer, Rene, 157
Meyers, Mark B. 106
Michal, Gary, *76*, 120
Miller, Alan K., 137, 141-144,
Miller, Edward, 50
Miller, Robert D., 215
Mills, Michael J., 33, *68*, 107, 117-118, 121
Mitchell, Kim, *116*, 117
Moberly, Warren, *118*, 121,
Morgan, Eric, 72
Morris, Bill, 204
Morris, Jeffrey, 212
Morse, Samuel, 27
Murzik, Mike, *116*

Nakamura, Shuji, 50
Narasimhan, Subha, *87*, 110,
Nabarro, Frank R. N., 99-100, 185
Nason, Don, 119
Nastasi, Michael, 189
Newsom, John Flesher, 4
Newsom, John Branner, 4
Nicolet, Marc, 172
Nieh, T. G., *68*, 107, 158
Niewenhuizen, Rob Van den, 143-144
Nishi, Yoshio, 190
Nix, William Dale, 33, 46-47, 57, 65-70, 74, 76, 80, 83-84, 93, 99, 103-104, 107-109, 110, 112-114, 116-118, 121, 124-125, 127-128, 175-178, 189
Norton, F. H. 138
Noufi, Rommel, 157

Obama, Barack H. (President) 90
O'Hara, Sydney, 58
O'Hayre, Ryan, 107, 155
Oldberg, Terry, 142
Olgivie, Robert E. 52
Oliver, Warren C., 69, 108, 120, *121*, 165
Olson, Greg, 173
Ong, Marcus, *230*
Ono, Kanji, 73
Orgutani, Tarik Ömer, 72, 123, *124*
Oshman, Kenneth, 139
Oster, George, 156
Öveçoğlu, Mustafa Lütfi, 124
Oyama, Toshi, *118*, 120, *121*, *122*
Ozguven, Nevran, 124
Özkan, Sinan Cengiz, 124

Pablo, Juan de, 156
Paine, David, 165
Paine, Thomas O. 30-32
Pankove, Jacques, 50
Panofsky, Wolfgang (Pief), 87
Pantell, Richard, 139
Park, Charles Frederick, Jr. 36
Parker, Earl, 106
Parkman, Ralph, 30-33
Parlee, N. D. A. 124
Patel, Jamshed R. 157
Pauling, Linus, 150
Pease, Fabian, *148*
Persson, Kristin, 157
Peterson Laboratory, 18, 40, 51, 53-54, 87, 112, 116, 127, 135, 139
Peterson, Thomas F., 18, 54
Peterson, Mrs. Thomas F., 64
Pethica, John (Sir), 69
Pettit, Joseph M. 34, 38, 54, 59, 193
Peumans, Peter, 201
Pharr, George, 69, *76*, 107, 108, 117, 120
Pinsky, Paul, *120*
Pizzo, Patrick P. P., 33, 120
Plummer, James D. (Jim) 163-165, 167, 173, 181
Polman, Alberto, 196
Poo, Mu-ming, 214
Poppa, Helmut R. 157
Pound, Guy Marshall, 57, 60-61, 71, 75-78, 80, 86, 94, 99, 107, 163, 191, 229
Prinz, Friedrich B. (Fritz) 41, 145, 153-155, 181,
Proft, Jim, *118*,

Quate, Calvin, 139
Queisser, Hans-Joachim, 100

Radcliffe, S. V. 106
Radjy, Shahla, 110
Ralph, Brian, 132
Raistrick, Ian, 41-42
Ralls, Ken, *56*
Ramanathan, Shriram, 106
Rau, Charles A. *74*, 78, 79
Read, T. A. 99
Read, Thomas L. 80, *119*, *120*,
Redfield, David, 64, 157
Reed, Evan J. 134, 202, 223-224
Reed, Leonard, 55, 100
Reiley, Tim, *120*
Reynolds, Richard A., 50
Rhines, Walden C. 51
Richter, Burton, 146
Rickey, Dave, *121*, *122*
Ritchie, Robert, 184, 185
Robertson, Diane (Margel), 110
Robinson, Arthur L., 63, 64
Robinson, Heywood, 121
Robinson, Steven, *115*, *116*, 117
Ross, Bernard, 78
Roth, John, 173
Rothman, Steven J., 31
Roy, Rustum, 106
Ruano, Oscar, *68*, *121*
Rudee, M. Lee, 34
Rutter, J. W., 59
Ryan, Harris J., 15

Salleo, Alberto, 175, 200-204
Sands, Timothy, 202
Saraswat, Krishna, 163, 164, 165, 190,
Sarnoff, David, 50, 64 (Labs)
Sauveur, Albert, 5, 21-22, 27
Saxton, Harry, 84
Scattergood, Ronald, 95
Schalliol, Willis L., 31
Schellinger, A. Kenneth, *22*, 33
Scherman, Oren A. 215
Schieber, Jay, 156
Schiffer, Menahem Max, 92
Schlom, Darrell G. 107, 108, 153
Schmidt, Chuck, 120
Schöck, Professor, 154
Schoen, David T. 80
Scholl, Jonathan, 222
Schön, Hendrik, 198
Schultz, George, 118
Schultz, Jerold, 99
Schulz, Max J. 100
Seo, Kang-Ill, *125*
Shaw, John, 139
Shen, Z.-X. 161, 198
Shepard, Orson Cutler, 7, 12-13, 15-17, 21-22, 25-36, 38, 45, 47, 53, *56*, 57, 59, 67, 72, 97, 123, 165, 175-177
Sher, Arden, 157
Sher, Renee, 174
Sherby, Oleg Dimitri (Sherbynin), 33-34, 37-38, 45-49, *68*, 72, 76, 80, 83, 99, 107, 108, 113-114, 115-117, 124, 142=143, 158, 184, 189-191
Shikamura, Takeo, 228
Shin, Dong-Woon, *125*
Shinde, Subhash, *118*
Shirikawa, Hideki, 194
Shirley, David, 168, 169
Shultz, Tyler, 183
Shyne, John Cornelius (Jack), 57, 61, 65, 71-73, 77-78, 175, *177*, 184
Siegbahn, Kai, 161, 168
Siegman, Anthony, 139
Sinclair, Robert, 74, *76*, 84, 103-104, 106, 121, 122, 128, 129, 132-136, 163, 164, 165, 166, 170, 171, 173, 175, 178, 181, 223, 229
Sinnott, Maurice J. 72
Smallman, Ray, 38, 98,
Smith, Donald L., 168
Smith, J.P., 4
Smith, Jim, *121*
Smith, John, 172
Spaepen, Frans, 172
Spakowitz, Andrew, 145, 155-156, 196, 201, 213, 214,
Spicer, William Edward, 57, 62, 87, 145-147, 152, 168, 169
Spingarn, Jay, *76*, 121
Sprague, B. Sheldon, 106
Stanford, Thomas Welton, 19
Stanford, Welton, 19
Starr, Chauncy, 78
Stebbins, Horacio W., 16
Stephens, John, *68*, *93*, *117*, 118, *121*, 122

Stein, Richard S. 106
Steinbeck, John, 66
Sterling, J. E. Wallace, 38
Stern, Dan, 120, *122*,
Sternstein, Sanford S. 106
Stevenson, David Austin, 37, 38, 45, 49-52, 115-117, 124
Stoughton, Bradley, 28
Stoughton Award 107, 166
Street, Robert, 203
Streiffer, Stephen, 153
Stringer, John, 107, 157,
Stringfellow, Gerald B., 63, 64, 108
Stroh, A. N. 94
Stucky, Galen, 194, 197,
Stuke, Josef, 100
Sullivan, Walter, 48
Suo, Zhigang 95
Surek, Tom, 119, *120*
Swanger, Lee A. 80, 94

Taleff, Erik, *68*
Taratorin, Alex, 182
Tarshis, Lemuel, 58, 59
Tasooji, Amaneh, 141
Terman, Frederick E., 17, 35-40, 54, 55, 58, 124, 167, 181, 184, 197, 221, 225
Terman, Sibyl, 35
Tetelman, Alan Stephen, 57, 71-74, 77-79, 81, 84, 91-93, 99, 106, 184
Thomas, G. 133
Thomas, Mike, *116*, 117
Toney, Michael, 201, 207, 208,
Tickell, Frederick G., 14, 16, 21
Tietjen, James J., 50, 106
Tiller, William Arthur, 34, 57-62, 71, 72, 77, 86, 99, 152, 174, 175, *176*, *178*, 191,
Tiner, Nevzat, 29, 123
Tresidder, Donald B., 16
Triplet, Baylor, 157
Turnbull, David, 38, 60, 86, 90, 136, 172

Uchic, Mike, *68*, 70,
Udin, Harry, 35

VanderPlas, Hugh, *116*, 117
Vinci, Richard P. 107, 174

Wachsman, Eric, 108
Wadsworth, Jeffrey, 46, 49, *68*, *76*, 107, 117, 120, 157, 158-159
Wagner, Christian N. J., 73
Wagner, Karl, 40
Wallace, Richard, 211
Wang, Connie, 153
Wang, Shan X. 136, 147, 149, 150, 181-184, 229
Wang, Zhen-Gang, 156,
Ward, Robert G. 98
Warshaw, I. 106
Weertman, Johannes, 38, *94*
Weihs, Tim, 121
Weinberg, Leonard R. 106
Wen, John, 41-43
Weppner, Werner, 41-42
West, Anton J. "Bud," 80, 84, *119*,
Westman, Russell A. 73
Whalen, Rob, *120*
White, Rick, *76*, 117, 120,
White, Robert L., 57, 62, 138, 139, 145, 147-150, 152, 182, 183,
White, Robert M., 157, 182,
Whittingham, M. Stanley, 41, 108
Wickersheim, Kenneth A. 145
Widom, Jennifer, 184
Wilbur, Ray Lyman, 9, 11, 15-16
Wilson, Judge E. M., 21-22
Willis, Bailey, 9, 11, 27,
Wing, Charles B., 15
Wise, Henry, 157
Wolf, Richard C., 53
Wong, Simon, 164
Wright, Tim, *120*
Wright, Wendelin, 103-104, 157, 166-167, 174
Wu, Hsaio, 228
Wulff, John 35, 49

Yamaoka, Tetsuji, 213
Yaney, Debbie, 113
Ye, Ziaofei, 210
Yoo, Ji-Beom, 125

Zednik, Ricardo, 80
Zimmerman, Roy, 67

Made in the USA
Lexington, KY
26 October 2019